PWA90

A LIFETIME OF EMERGENCE

PWA90

A LIFETIME OF EMERGENCE

Editors

Premi Chandra (Rutgers University)

Piers Coleman (Rutgers University)

Gabi Kotliar (Rutgers University)

Phuan Ong (Princeton University)

Daniel L Stein (New York University)

Clare Yu (University of California, Irvine)

World Scientific

NEW JERSEY · LONDON · SINGAPORE · BEIJING · SHANGHAI · HONG KONG · TAIPEI · CHENNAI · TOKYO

Published by

World Scientific Publishing Co. Pte. Ltd.

5 Toh Tuck Link, Singapore 596224

USA office: 27 Warren Street, Suite 401-402, Hackensack, NJ 07601

UK office: 57 Shelton Street, Covent Garden, London WC2H 9HE

Library of Congress Cataloging-in-Publication Data

Names: Chandra, Premi, editor. | Coleman, Piers, 1958–   editor. | Kotliar, B. Gabriel, editor. |
    Ong, N. Phuan, 1948–   editor. | Stein, Daniel L., editor. | Yu, Clare C., editor. |
    Anderson, P. W. (Philip W.), 1923–   honoree.
Title: PWA90 : a lifetime of emergence / edited by Premi Chandra (Rutgers University),
    Piers Coleman (Rutgers University), Gabi Kotliar (Rutgers University), Phuan Ong
    (Princeton University), Daniel L. Stein (Courant Institute, NYU), Clare Yu (UC, Irvine).
Other titles: Philip W. Anderson 90
Description: Singapore ; Hackensack, New Jersey : World Scientific Publishing Co. Pte. Ltd., [2016] |
    2016 | Includes bibliographical references.
Identifiers: LCCN 2015038645 | ISBN 9789814733618 (hardcover ; alk. paper) |
    ISBN 981473361X (hardcover ; alk. paper) | ISBN 9789814733625 (pbk ; alk. paper) |
    ISBN 9814733628 (pbk ; alk. paper)
Subjects: LCSH: Anderson, P. W. (Philip W.), 1923– | Condensed matter. | Quantum theory.
Classification: LCC QC173.454 P93 2016 | DDC 530.12--dc23
LC record available at http://lccn.loc.gov/2015038645

British Library Cataloguing-in-Publication Data
A catalogue record for this book is available from the British Library.

Contents

Preface

"Curiosity is, in great and generous minds, the first passion..."

— Samuel Johnson

In December 2013 a community of physicists gathered in Princeton on the occasion of Phil Anderson's 90th birthday to celebrate his curiosity and achievement across a remarkable career spanning more than six decades. As codified in his oft-quoted phrase "More is Different", Phil has been the most forceful and persuasive proponent of the radical (in the 1970s), but now ubiquitous, viewpoint of emergent phenomena: truly fundamental concepts can and do emerge from investigations of Nature at each level of complexity or energy scale. The workshop's title, "PWA90: Emergent Frontiers of Condensed Matter", was thus inspired by Phil's ideas of emergence that have deeply influenced developments in their original realm of condensed matter physics as well as in high energy physics, astrophysics, economics, computational optimization and computer science. Phil's insights have had a profound impact on such a broad number of topics that a two-day symposium could only touch on some of them. The meeting had six scientific sessions and two panel discussions; the program is included here as an Appendix. Each talk had a significant amount of time allocated for discussion that often was quite lively. Phil sat in the front row of the auditorium for the entire meeting, and asked questions after almost all the lectures. True to form, he enjoyed some healthy differences of opinion with some of the speakers. The meeting was well attended by current students, mainly from the tri-state area, who were inspired by the presentations and the animated discussions.

This workshop brought together many of Phil's students, postdocs, collaborators and colleagues from throughout his career, and some personal reminiscences from the celebration dinner are included in this volume. The scientific talks at the meeting covered several representatative topics that

Phil has deeply influenced. Let us now make Phil blush: the seminal work associated with his name includes Anderson localization, the Anderson local moment model of magnetism, the Anderson–Higgs mechanism, the Edwards–Anderson order parameter in spin glasses, the Anderson–Brinkman–Morel (ABM) phase of superfluid ^{3}He, Anderson's dirty superconductor theorem, the Anderson–Kim model of flux creep in superconductors, and the Anderson–Holstein impurity model and we are still not done! Furthermore many of the profound concepts that Phil has introduced and/or developed include broken symmetry and rigidity, superexchange, spin liquids, scaling in interacting systems, frustration and local moments, all of which are now part of the standard lexicon of the field. The challenges of celebrating Phil's deep and broad influences within the span of a two-day meeting should now be clear to all. The scientific talks at this workshop reviewed Phil's crucial contributions to representative areas and emphasized subsequent developments; the topics covered were Anderson localization, the Anderson–Higgs effect, frustrated magnetism and heavy fermions, high-temperature superconductivity, superfluids and entanglement, biological physics and neutron stars.

On the first day, the session on Anderson localization combined a historical introduction by Elihu Abrahams (UCLA) with a forward-looking discussion of modern challenges of many-body localization by Boris Altshuler (Columbia). The session on the Anderson–Higgs effect included presentations by Edward Witten (IAS) and Frank Wilczek (MIT) on Anderson's contributions from the perspective of high-energy physics. In the session on frustration and heavy fermions, Patrick Lee (MIT) discussed Anderson's theory of resonating valence bonds, and Ali Yazdani (Princeton) described scanning tunnelling microscopy as a probe of heavy fermion materials. The first day ended with a very animated panel discussion involving Bill Brinkman (Former Director, Office of Science, Department of Energy) and Anderson's students Pierre Morel, Ted Hsu, Joe Wheatley and Joe Zhou, who described how physics helped them in non-academic careers that included science policy, politics and investment banking.

The second day began with a session on strongly correlated superconductors where Maurice Rice (ETH, Zurich), Gabi Kotliar (Rutgers), J. C. Campuzano (UIC) and Mohit Randeria (Ohio State) discussed current progress towards a collective understanding of high-temperature superconductivity. The second session on superfluids, entanglement and biology involved talks by Albert Libchaber (Rockefeller) on the problem of self-reproduction, by Anthony Leggett (UIUC) on the concept of superconducting liquids and Duncan Haldane (Princeton) on topologically entangled matter. In the final session, Mal Ruderman (Columbia) spoke about neutron stars. Then there was a lively description by Walter Kohn (UCSB) of Phil when both he and Walter were graduate students at Harvard. The symposium ended with a panel discussion on historical and grand challenges of condensed matter physics with participants Erio Tosatti (Sissa, Italy), David Thouless (Washington) and T. V. Ramakrishnan (Indian Institute of Science, Bangalore).

We hope that this volume, which includes written versions of many of the symposium talks and some personal reminiscences, will serve to inspire many present and future students and practitioners in the area of Emergent Condensed Matter Physics. We note that Ted Geballe (Stanford), and Scott Kirpatrick (Hebrew Univ. Jerusalem) were unable to attend the meeting but generously contributed articles to these Proceedings.

We wish to thank all the speakers, contributors, and participants for their roles in making the workshop such an inspiring, enjoyable and memorable affair. In particular, we express our deep gratitude to Catherine Brosowsky and the staff at Princeton University for making all the detailed arrangements towards running the workshop smoothly, and to Fran de Lucia at Rutgers University for processing the graduate student expenses. For their generous support of the conference PWA90, we wish to thank the National Science Foundation (NSF grant PHY-1401789), the Gordon and Betty Moore Foundation, the Department of Physics at Princeton University, and the Princeton Center for Complex Materials. We are grateful for the hospitality of the Aspen Center for Physics (supported by National Science Foundation grant PHY-1066293) where some of us (CCY, PC, PC, and GK) put the finishing touches on these proceedings.

Finally, a toast. On behalf of the physics community past, present and future, we raise a glass to Phil and express our heartfelt thanks for a cornucopia of discoveries and insights that have provided the foundation for the work of generations to come. Isaac Newton once said, "If I have seen further than others, it is by standing on the shoulders of giants". Thank you, Phil, for giving us shoulders to stand on...and we will do our very best to be curious and to keep exploring!

Premi Chandra
Piers Coleman
September 2015   Gabi Kotliar
Phuan Ong
Dan Stein
Clare Yu

Phil Anderson offers a comment to one of the speakers. Background: Phuan Ong.

Phil Anderson poses a question to one of the speakers. Foreground: Elihu Abrahams.
Background: Susan Anderson.

Conference photograph

Phil Anderson cuts the birthday cake.

Bill Brinkman and Maurice Rice

Jim Langer and David Thouless

Walter Kohn, Daniel Tsui, Douglas Osheroff, Phil Anderson, Frank Wilczek and Edward Witten

Gabriel Kotliar, Phil Anderson, Clare Yu and Piers Coleman

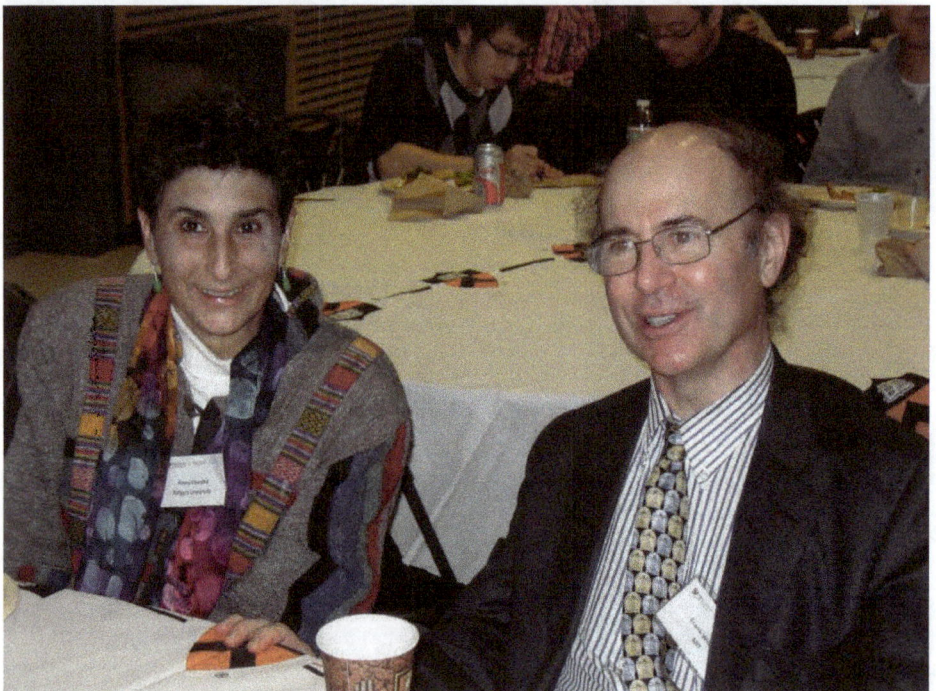

Premi Chandra and Frank Wilczek

Ali Alpar and Jim Sauls

Bert Halperin and Walter Kohn

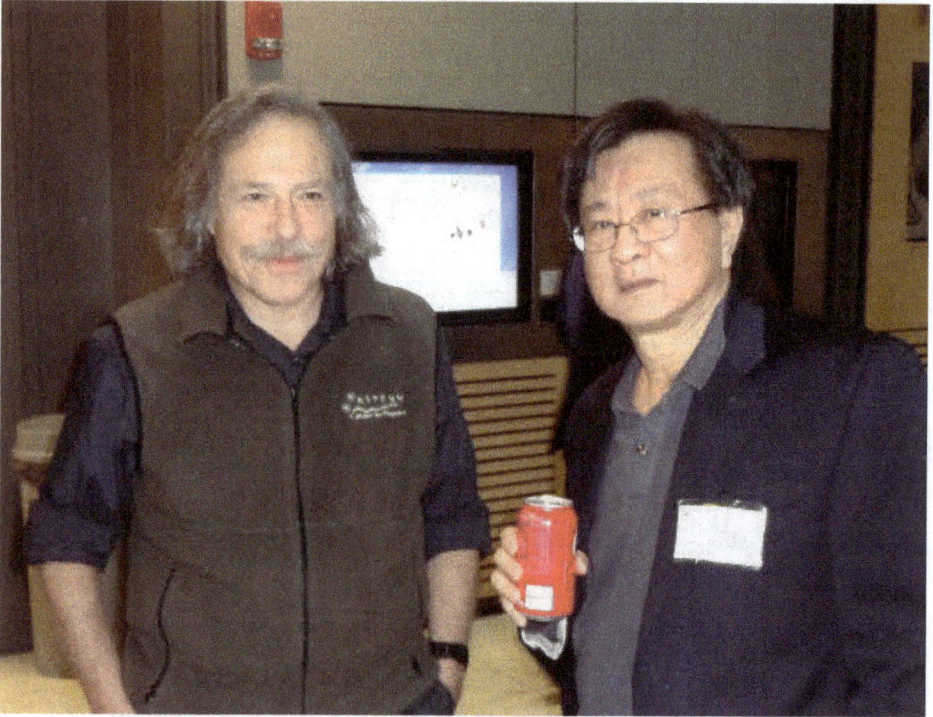

Paul Chaikin and Phuan Ong

Walter Kohn and Phil Anderson

Frank Wilczek, Phil Anderson and Erio Tosatti

Don Hamann, Patrick Lee and Phil Anderson

Michael Gershenson and Boris Altshuler

Duncan Haldane and David Vanderbilt

Sriram Shastry and Zahid Hasan

Phil Anderson, Ganapathy Baskaran and Krastan Blagoev

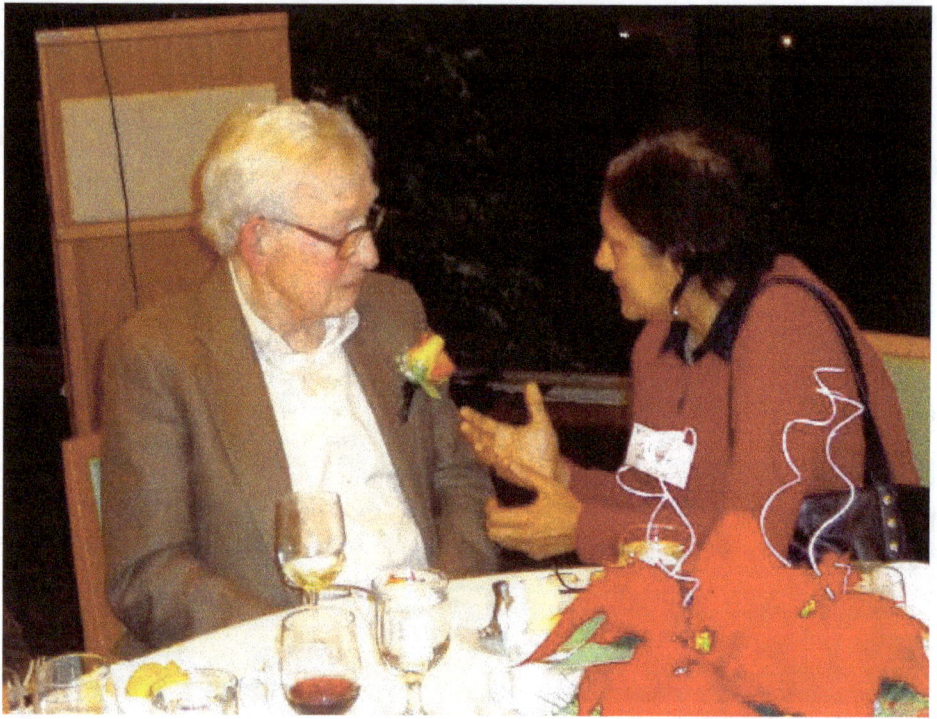

Phil Anderson and Nandini Trivedi

Maurice Rice and T. V. Ramakrishnan

Myriam Sarachik, Dieter Vollhardt and Erio Tosatti

Alexei Tsvelik and Duncan Haldane

Frank Wilczek and Joe (Zhou) Zou

Edward Witten and Eva Andrei

Pierre Morel and David Thouless

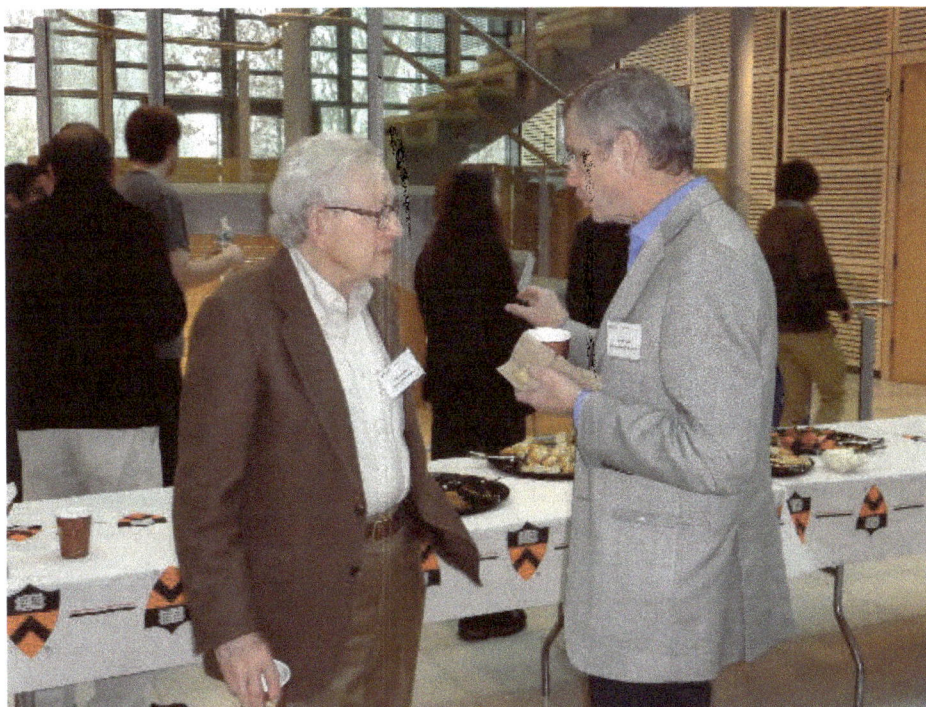

Phil Anderson and Jim Sauls

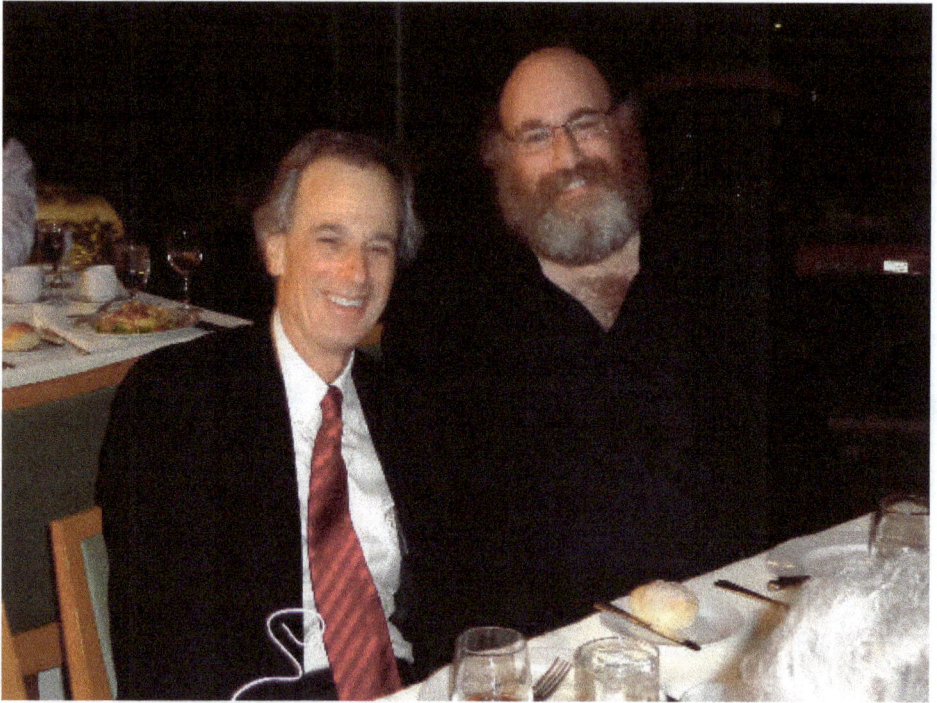

Daniel Stein and Bill Bialek

Phil Anderson with his daughter, Susan Anderson

Bert Halperin and Daniel Fisher

Pierre Hohenberg and Kathryn Levin

Recollections of a Graduate Student

Khandker A. Muttalib

Department of Physics, University of Florida
Gainesville, FL 32611-8440, USA
muttalib@phys.ufl.edu

These are some recollections of my graduate years at Princeton in the late seventies and early eighties under the supervision of Phil Anderson.

Some Recollections

I came to Princeton in the year 1977, all the way from Bangladesh, to study High Energy Physics. There was no course requirement in the graduate program, so many of us started studying for Prelims, the first big obstacle towards a Princeton degree, and I also enrolled in some courses relevant for high energy theory. When news broke sometime in the Fall that one of our faculty members, not in high energy, just received the physics Nobel Prize, I was curious to understand what it was all about. A lecture was hastily arranged by the department. Listening to Philip Anderson, although I did not understand even some of the basics, I realized for the first time that condensed matter could be more exciting than what I was exposed to in my undergraduate years.

After the Prelims, the next hurdle was the Qualifying exams. It basically required us to choose a standard textbook in each of the four major areas in physics and teach them to ourselves. Thus my only training in condensed matter physics, up until the summer of 1979, was the several intense weeks spent on Ashcroft and Mermin.[1] It was then quite a series of coincidences that when I was beginning to decide on a topic for my thesis, the famous "scaling theory of localization" by Abrahams, Anderson, Licciardello and Ramakrishnan[2] came out in March 1979, Ramakrishnan came from India to visit Princeton for an extended period of time, and I ended up spending a summer working on a project with him.

It was Ramakrishnan who encouraged me to try out a research career in condensed matter. I was deeply hesitant, knowing my non-existent background, but he assured me that it was the perfect time to approach Phil Anderson to work on some aspect of the newly discovered scaling theory of localization. I am not sure how I gathered the courage (a lot of which I am sure I borrowed from Ramakrisnan), but eventually one day I found myself in Phil's office asking if he would give me a chance. I did not realize at the time how nervous I must have been, but it was obvious to me afterward because I could not recollect any question that Phil must have asked me. All I remember is that he sent me to Bell Labs to work with Patrick Lee for a brief period of time and then come back and report to him. I later realized how gracious and thoughtful it was on Phil's part to basically give me some time to gather a little more confidence, when it was more natural for him not to "waste" time on a nervous student from a very different culture.

On the first day after I came back from Bell Labs, Phil stood in front of the blackboard in his office and wrote down five possible research questions I could choose from. I wrote them down diligently, since they were all on topics I had no idea about. I took a few months to search the literature trying to figure out what Phil's questions really meant. I decided that the topics ranged over such a widely disjoint spectrum of condensed matter that with my background, I had no hope of understanding what the problems were in more than one area in any reasonable time. Six months later, I went back to report that I have decided to work on the effects of disorder on superconductivity. However, I must have said something, because he shook his head, went over to the blackboard, and gave me a two minute lecture on what he really meant. It was another six months before I would come back to tell him that now at least I understood the problem. When I think about it now, it seems amazing that Phil would be so patient with a student, given his general impatience with mediocrity and incompetence.

Once the problem was decided, namely to consider the effect of the anomalous diffusion of electrons on disordered superconductors (Phil expected an increased electron-phonon interaction due to the slowing down of the electrons) I wrote down the most general Hamiltonian and suggested that I perturbatively try to calculate a particular set of diagrams. Phil closed his eyes for several minutes and then proclaimed that my calculation will give me zero effect in the end. I was happy to have come up with something on my own, and I could not find any symmetry reason for the result to be zero, so I took a month calculating the diagrams anyway, which eventually gave me a negligible effect. Next time when Phil made a similar comment on another of my proposals, I gathered my courage and asked, "could you please think aloud?" Phil looked at me for some time, and then perhaps considering it as an amusing exercise, tried "thinking aloud". At that point I realized that while I was always thinking in terms of the Hamiltonian and minimum energy state of the system, Phil had certain aspects of the entire interacting many-body wave function in his head and he was basically telling me that the overlap of the relevant wave functions in my case is small and therefore the effect is expected to be negligible. It was a rather frustrating experience, because it was clear to me that the art of forming such pictures in your head is not something that can be either taught or learnt. Nevertheless, I would often ask Phil to "think aloud", and while I was not able to use any of the insights to solve my problem, they undoubtedly saved me from a host of dead-ends in my beginning years.

My results on the electron-phonon interactions disappointed Phil. It was a null result, and we never published it.[3] I was not expecting him to give it up so easily, but he believed in my reasoning (he must have updated his initial optimistic "pictures") and asked me to move on. The idea was that the impurities move with the lattice, which neutralizes the effect on electron-phonon interactions that Phil was hoping for; but then it might still be worth looking at the electron-electron interactions. It was at this time that I obtained some unexpected results. It turned out that the Coulomb interaction, which appears only as a pseudo-potential due to slow-moving phonons, nevertheless can increase with increasing disorder essentially due to the anomalous diffusion of the electrons. This affects the superconducting transition temperature quite dramatically, and we showed that they were especially relevant for the "high-T_c superconductors" of the time, the A-15 compounds, with T_c of the order of 20K.

When I gave a talk on my results at Bell Labs, many in the audience objected that my final conclusion violates the Anderson Theorem (that non-magnetic impurities that do not break time-reversal symmetry do not affect normal superconductors) and therefore I must have made some blatant mistake. I am not sure how I would have handled such criticism coming from some of the biggest names in condensed matter physics. But Phil stood up and simply said, "I know that this does not violate the Anderson Theorem". After that, more serious discussion followed.

I used to visit Phil once every year for quite some time after I left Princeton, and each time he would take the time to learn about all that I had done in the meantime. After my graduation I had given him a small souvenir from Bangladesh, made of buffalo horn; each time I visited him he would point it out on his shelf and say that he still had it, just to make me happy. He advised me to accept an invitation from Doug Stone to join Yale as a postdoc (even after two postdocs at Chicago and Brookhaven), and also later to accept an offer from Peter Wölfle to join the University of Florida as an Assistant Professor, because he had great respect for both of them. He visited University of Florida and gave a talk and my wife Flora had the opportunity to cook dinner for him. He was quite impressed with Bangladeshi food, and after returning back wrote a beautiful letter to Flora appreciating her cooking.

Anderson's own works speak for themselves, and there is no need for me to add anything to that. However, I guess that less is known about the person. It is true that I missed a lot of the get-togethers he and his wife Joyce arranged for the students in their house because I was staying in New York where Flora was doing her PhD in physics at NYU. Towards the end of my graduate studies I only saw him (or sometimes just talked over the telephone) when I was stuck and there was critical need for his insight. But even then I feel that I have been really fortunate to know the person quite well. I am indeed privileged to have the opportunity on the occasion of his 90th birthday to recount my association with him, and tell the world how extraordinarily gracious, thoughtful and supportive he has been to me. I have heard the same story from many of his students.

References

1. N.W. Ashcroft and N.D. Mermin, *Solid State Physics* (Holt, Rinehart and Winston, 1976).
2. E. Abrahams, P.W. Anderson, D.C. Licciardello and T.V. Ramakrishnan, *Phys. Rev. Lett.* **42**, 673 (1979).
3. K.A. Muttalib, *Strongly Disordered Superconductors*, PhD Thesis, Princeton University (1982).

P. W. Anderson Seen Through the Eyes of a Student

Clare C. Yu

Department of Physics and Astronomy
University of California, Irvine
Irvine, CA 92697, USA
cyu@uci.edu

These remarks were given after dinner at PWA90. A former student of P. W. Anderson reminisces on what it was like to work for Phil by relating some anecdotes.

Some Anecdotes

I was both an undergraduate and graduate physics student at Princeton. I first heard about Phil when I was a junior. One morning in October 1977, I was sitting in a classroom, waiting for my statistical mechanics class to begin, when the professor, Tom Carver, came bounding into the room and exclaimed, "We're all very excited! Phil Anderson has just won the Nobel Prize!" I thought, "Who? What did you say his name was? Why are we excited? Is he a professor here?" I was, of course, to destined to become much more acquainted with Phil.

Phil worked primarily in solid state physics. As a junior, this sounded to me like a very dull subject. I pictured a solid block of aluminum just sitting on a table, not doing anything. Boring! But my view soon changed in the summer following my junior year when I read a *Scientific American* article[1] about the discovery of superfluid ^{3}He. So in the fall of my senior year, I asked Phil to be my senior thesis advisor and told him that I thought superfluid ^{3}He was interesting. He said that was too hard for a beginning student and suggested I work on solid ^{3}He, which I did. He captured my interest when he pointed out that even though the helium atom is very simple, a collection of them produces unexpected phenomena such as superfluidity and magnetic phases. Basically he was giving me an example of "More is Different"[2] without my realizing it.

In the course of working on my senior thesis, Phil taught me the art of making estimates and plugging in numbers. He explained that you do not just plug numbers in to a calculator; you think about each number and physical constant. You keep the important physical constants in your head and you use units that are suitable for the problem, e.g., do not use Joules/mole, use Kelvin/atom. That was an important lesson that I now try to pass onto those working for me.

I stayed on at Princeton as a graduate student. My dream was to do quantum gravity. Doug Osheroff, my Bell Laboratories Graduate Research Program for Women mentor, did not think that quantum gravity was a good career choice since it would be very difficult to get a job in that field. In fact, he thought it was a mistake and told me so. I figured, "He's an experimentalist. What does he know about theory?" The real problem was that Doug was not telling me what I wanted to hear, so I ignored him. (He did not have a Nobel Prize yet.) Realizing that his good advice was falling on deaf ears, Doug went and told Phil. Phil called me into his office, sat me down, and managed to knock some sense into me. So I decided to do condensed matter theory instead and work for Phil.

For my thesis, Phil suggested that I prove it was impossible to have high-temperature superconductivity. (This was in 1981 before the discovery of high-temperature superconductivity.) Fortunately, that did not last long and I moved onto other topics.

In the course of working with Phil, there were times when he would get frustrated with me as one does with students from time to time. Once, in complete exasperation, he said, "Theoretical physics isn't doing calculations." I said nothing, but thought, "Ok, so what is it?" Reading my mind, he went on to say, "It's setting up the problem so that any fool could do the calculation."

Communicating with Phil was a challenge. For example, I learned to summarize everything in 90 seconds or less, or else he would start to read his mail or fall asleep. Also his words had a different meaning when he used them than when other people used those same words. So I would read his papers to try to figure out what he meant by those words. When he talked about "it", I often wonder what "it" was. As he spoke, I would build a picture in my mind and then suddenly come to my senses and realize that the picture made no sense. So I would stand in his office stammering "What?" When I left his office, I would try to remember all the words he used, and put them together in all possible combinations to try to find one that made sense. That is how Phil taught me to think.

As a graduate student, I had the privilege of editing Phil's book, *Basic Notions of Condensed Matter Physics*.[3] I would try to reword things to make them clearer. However, sometimes there were sections that I could not understand. So I would ask others like Mike Cross or Sue Coppersmith for help. But I remember there was one particular passage which no one could comprehend. So I showed it to Phil. He looked at it and said, "I guess I wasn't very clear." He stared at it for a while longer, then finally understood what he meant, and explained to me. Then he looked back at the passage and said, "I think I said it ok." But added, "But you can rewrite it if you want," which I did.

I would like to end with one final anecdote that occurred after I graduated. At the 1987 American Physical Society March meeting in New York City, Phil was giving at talk at the symposium on the History of Super-conductivity. The title of Phil's talk was "It's Not Over 'Till the Fat Lady Sings." A lady at the convention center told us that the saying was a reference to the singer, Kate Smith, who used to sing "God Bless America" at the end of the Philadelphia Flyers' hockey games.[a] Another former student of Phil's, Yaotian Fu, and I thought it would bring down the house if we could get a fat lady to sing "God Bless America" at the end of Phil's talk. Since we were in New York City, we figured we could hire a singer who would fit the bill. So we called a local talent agent who told us that it would cost $200 if he could even find such a singer on such short notice. In 1987, $200 was a lot of money for two postdocs, so we decided to solicit contributions. If ten people each gave $20, we would have it made. So we told the agent to line up a singer and we started to hit up our friends for money. I asked my friend Sue Coppersmith who agreed to not only contribute, but to help us work the floor. She said when she asked Ted Geballe, a long time friend of Phil's, he whipped out his wallet and asked, "How much do you need?" Sue responded that we were not accepting any money until we actually knew someone had agreed to sing. Unfortunately, the talent agent was not able to find a singer, and Phil gave his talk with no surprise endings. But word of our scheme spread and Sue said that people she did not even know came up to her the next day and offered to contribute. I never told Phil until years later at an event like this one.

---

[a]This is probably not the origin of the saying since Kate Smith sang at the beginning of the Flyers' games.

In closing, I would like to thank Phil for being my advisor and wish him a very happy birthday.

References

1. D.M. Lee and N.D. Mermin, Superfluid helium 3, *Sci. Am.* **235**, 56–60, 62, 64, 67–68, 70–71 (1976).
2. P.W. Anderson, More is different, *Science* **177**(4047), 393–396 (1972).
3. P.W. Anderson, *Basic Notions in Condensed Matter Physics* (Benjamin/Cummings, Menlo Park, CA, 1984).

Random Walks in Anderson's Garden: A Journey from Cuprates to Cooper Pair Insulators and Beyond

G. Baskaran

The Institute of Mathematical Sciences
C.I.T. Campus, Chennai 600 113, India and
Perimeter Institute for Theoretical Physics
Waterloo, ON, N2L 2Y6 Canada
baskaran@imsc.res.in

Anderson's Garden is a drawing presented to Philip W. Anderson on the eve of his 60th birthday celebration, in 1983, by a colleague (author unknown). This cartoon (Fig. 1) succinctly depicts some of Anderson's pre-1983 works. As an avid reader of Anderson's papers, a random walk in Anderson's garden had become a part of my routine since graduate school days. This was of immense help and prepared me for a wonderful collaboration with Anderson on the theory of high-T_c cuprates and quantum spin liquids at Princeton. Here I narrate this story, ending with a brief summary of my ongoing theoretical efforts to extend Anderson's RVB theory for superconductivity to encompass the recently observed high-temperature ($T_c \sim 203$K) superconductivity in solid H_2S at pressure $\sim$200 GPa. In H_2S molecule, four valence electrons form two saturated covalent bonds, H-S-H. These bond singlets are *confined Cooper pairs* close to chemical potential. Solid H_2S is a *Cooper pair insulator*. Pressure changes the structure and not the number of valence electrons. Bond singlet pairing tendency continues and new S-S and H-H bonds are formed. S-S bonds are mostly saturated. However, hydrogen sublattice has unsaturated H-H bonds. It prepares ground for a RVB superconducting state.

1. Introduction

I visited Phil Anderson at Princeton over the three-year period 1984–1987. An intense collaboration with Anderson, or rather *a resonance* took place

Fig. 1.   *Anderson's Garden*: A drawing presented (author unknown) to P. W. Anderson on his 60th birthday (1983). Anderson localization, local moment formation, motional narrowing, neutron stars are visible among other things.

for a few months between November 1986 and March 1987, during which I was drawn deeply into the world of strongly correlated electron systems, novel quantum phases and high-T_c superconductors.[1] Interestingly, while at Princeton, before high-temperature superconductors appeared on the scene, I dabbled seriously with the idea of changing field into grey matter (neuro) science, an idea which Anderson supported. However, the enticing influence of new challenges from cuprates and Anderson's wholehearted involvement in the project changed my course.

The first part of this article is a personal account of my enjoyable stay at Princeton, my collaboration with Anderson and a brief account of how I got involved in the theory of quantum spin liquids and high-T_c superconductivity in cuprates. *Anderson's Garden* (Fig. 1) is a drawing presented to him on the eve of his 60th birthday, in 1983, by a colleague (author unknown). This cartoon (Fig. 1) depicts a thoughtful Anderson and his earlier works. As an avid reader of Anderson's papers, from my PhD days, random walk in Anderson's garden was my routine. This was of immense help and prepared me for a wonderful collaboration with Anderson on the theory of high-T_c cuprates and quantum spin liquids.

Over the years, I have suggested electron correlations and Anderson's RVB physics[2] to be present in normal and superconducting properties, to *varying degrees* in most of the new superconducting systems: fullerites, nickel borocarbides, Sr_2RuO_4 (p-wave superconductivity from strong correlation and Hund coupling), hydrated sodium cobalt oxide, MgB_2, ET and Bechgaard organic family, Boron doped diamond, iron arsenide family, doped graphene, doped $TiSe_2$, spin ladder compound and recently doped silicene and germanene.

The second part of this article summarizes my recent work, theory of superconductivity in the recently discovered superconductivity[3] in molecular solid H_2S with a $T_c \sim 190K$, under a pressure of 200 GPa. In H_2S molecule, two valence electron pairs form two saturated covalent bonds, H-S-H, and bind H and S atoms. We view paired valence electrons as *confined Cooper pairs* and molecular solid H_2S as a Cooper pair insulator. Pressure changes crystal structure, valence electron pairing pattern and deconfines some Cooper pairs: (i) sulfur atoms form a sublattice with saturated S-S covalent bonds, (ii) part of H atoms left behind in the small interstitials of the sulfur subsystem forms a dilated H-atom sublattice, a Mott insulator with unsaturated H-H covalent bonds in a resonating valence bond state and (iii) charge transfer between sulfur S and H subsystem, arising from a differing electronegativity dopes the Mott insulator and leads to superconductivity.

2. 1983–1984: Trieste to Princeton

My first meeting with Phil Anderson was at the International Centre for Theoretical Physics, Trieste, Italy in the summer of 1983. I was visiting ICTP and SISSA for an extended period, after failing to get a permanent academic job at my home country. Erio Tosatti, my wonderful host at Trieste, had invited Phil Anderson for a colloquium. After the colloquium was over, Erio came rushing. He said, "Phil is free, come and talk to him." Even though I had a great admiration for Anderson, I was reluctant and somewhat shy to meet him, because of his stature in the field. However, Erio insisted that I met. I agreed, as Arun Jayannavar, a good friend and visitor to ICTP, agreed to accompany me.

The post-lunch discussion with Phil lasted for more than an hour. Mostly I spoke. Anderson was in a sleepy/dreamy state; he made occasional remarks. I was describing my foray into CDW states, why supersolid ^{4}He should exist in spite of Anderson's criticism in his book, my variational approach to quantum roughening in solid ^{4}He and a few other topics that I was thinking about at that time. We parted. I was elated.

A continuing low job prospect back home and Erio's strong advice to cross the big ocean, forced me to try for visiting research-cum-teaching positions in the USA. I remember very well: at the end of a car ride near Trieste train station, Erio jotted down 13 names in a piece of paper. I wrote to all. Every one responded. The only affirmative response was from Phil Anderson. Actually he was apologetic that he could not fix a position for me at Bell Labs, as he had moved from Bell to Princeton. He wondered if I would accept a Visiting Research Staff (a visiting Assistant Professor position from Anderson's soft money) position at the Physics Department of Princeton University. I was overjoyed

In early September of 1984, I landed at Princeton with my wife and three little children. Later I learned from Erio that Anderson enjoyed my post-lunch discussion at ICTP. From Anderson I learned that Erio recommended me strongly. Erio, a good friend and admirer of Anderson was an earlier postdoc of him at Cambridge.

3. Random Walk in Anderson's Garden

Anderson's Garden is a wonderful drawing (Fig. 1) presented to Anderson on his 60th birthday, where Anderson's work is depicted as a nice garden. Gardener Anderson is wondering, "what shall I plant next?". It is a very imaginative picture with neutron star (neutron quake theory?) in the background, a buzzing bee showing motional narrowing theory in magnetic resonance, etc.

My random walk in Anderson's garden started during my PhD days, as a toddler. I was introduced to Anderson's works and helped to go into some depths by Rajaram Nityananda, N. Kumar, K. P. Sinha and (late) S. K. Rangarajan. Anderson localization theory, in the hands of Rajaram Nityananda and N. Kumar, was exposed to us from different angles. Further, I ran a journal club — most talks were on Anderson's papers, as and when they appeared in the journals. For example, I reviewed Edwards–Anderson spin glass model as the papers were appearing. Random walks in Anderson's garden became a habit.

Young Chandra Varma from Bell was a speaker at a TIFR summer school (1973) in IISc, Bangalore. He introduced us to Anderson lattice, Hubbard model, Mott insulators, heavy fermions, etc. Jayaraman of Bell Labs, who was on sabbatical at Bangalore setting up high pressure laboratory at the National Aeronautical Laboratory, also introduced us to the fascinating world of valence fluctuations, metal insulator transition in SmS, etc. Rajaram Nityananda, a fellow graduate student at that time, was ever ready to clear our doubts at the end of every lecture on any topic! N. Kumar, my mentor, got excited about topics in Mott transition, Jahn–Teller effect, Falicov–Kimball model, etc. I learned from C. N. R. Rao, who had just joined IISc, fresh experimental results on the enigmatic Mott insulator $LaCoO_3$, that exhibited low to high spin crossover and no long range magnetic order.

My supervisor K. P. Sinha introduced me to Anderson's 1958 paper on superexchange, in connection with a project he had suggested on doped EuO. He had stories to tell us about great physicists, whom he met during his stint at Bell Labs. Rajaram Nityananda had explained to me, adding his own insights, Anderson's Cargese lectures on local moment formation, etc. In 1977 H. R. Krishnamurthy arrived from Cornell and explained to us intricacies of Kondo phenomenon, valence fluctuation and how to understand them using quantum RG approach (standing on Anderson's poor man's scaling theory) that he, Wilkinson and Wilson had just developed. Anderson's masterly role in modern condensed matter physics was manifest. Strong correlation physics, including Mott insulator was in the air.

Other reason for a smooth entry to RVB theory is my long interest in quantum magnetism and help from friends. I had studied consequences of (Jordan–Wigner) Fermi sea in 1D spin-$\frac{1}{2}$ Heisenberg antiferromagnet, had used a projected wave function to go beyond spin-wave theory (a collaboration with Arun Jayannavar at ICTP), etc. Furthermore I had exposure to (i) Majumdar–Ghosh model and ordered valence bond states from J. Pasupathy and Sriram Shastry, (ii) anomalous $S(q, \omega)$ and quantum dynamics of 1D spin-$\frac{1}{2}$ Heisenberg antiferromagnet from Shastry, and (iii) Fazekas–Anderson work on RVB theory for 2D spin-$\frac{1}{2}$ antiferromagnet on the triangular lattice from (late) Patrick Fazekas himself, at ICTP. I vividly remember Patrick explaining Marshall sign convention difficulties on a triangular lattice, and a consequent *dynamic* sign convention, containing floating topological defects.

4. 1984–1987 at Princeton: Then Came Cuprates

My stay at Princeton was most enjoyable — talking to graduate students, Zhou Zou, Ted Hsu, Joe Wheatley, Yong Ren, Yaotian Fu, Mark Kwalay, postdocs Shoudan Liang, Anil Khurana, faculty, Phuan Ong, Dan Stein, Jim Sauls, Sajeev John, discussing with them a variety of modern and old topics. Discussion with Anderson was special. We provided mutual sympathy field and talked on topics that went beyond matter topics. For example, he used to tell low energy nuclear physics is incomplete using only two-body nuclear potentials; vacuum structure of QCD will have to play important roles, etc. During my stay, Anderson offered a graduate course on theoretical biology twice. Having had some exposure to Biology at Bangalore, I enjoyed both courses.

On the family front, we had a wonderful time at Princeton. The children enjoyed school. We had outing every weekend, covering nearby areas all the way up to Washington; and long drives to Niagara and Florida. Established lifelong friendship with wonderful people.

It was November 1986. End of nearly two and a half years of stay at Princeton, from September 1984, without any substantial achievement, except for few papers on cooperative ring exchange theory of fractional quantum Hall effect, spin glasses and traveling salesman problem. I had started feeling uneasy about my non-publishing mode. However, Anderson did not seem to mind. Once, in the middle of my stay, he made a comforting and prescient remark: "To do anything substantial it will take three years." He was kind and liked my natural exuberance, interaction with his students, visitors and others.

It was early November 1986. One early morning, Anderson came to my office and placed a photocopy of an article[1] on my table, and said, we should discuss this. The article was, a soon to become a Nobel Prize-winning one, by Bednorz and Müller in *Z. Phys.* It reported high-T_c superconductivity in cuprates. Anderson used to drive to Bell Labs to attend some talks. I have accompanied him sometimes. Anderson missed Kitazawa's talk at Bell Labs, where he introduced the exciting Bednorz–Müller discovery and its confirmation by Tanaka's group (in which he was a key member) and going beyond. A friend of Anderson had informed him about Kitazawa's talk and alerted him about high-T_c superconductivity in cuprates.

I read through the article of Bednorz and Müller carefully. Having accepted a job at the Institute of Mathematical Sciences, Chennai (then Madras), India and having decided to join in mid-1987, I was left with only about half year of stay at Princeton in Anderson's group. A wonderful opportunity to collaborate with Anderson has opened up. Secondly, having been introduced to Jahn–Teller effects in oxides from my PhD thesis advisor Prof K. P. Sinha and my mentor Prof N. Kumar at the Indian Institute of Science, Bangalore, I got excited about Jahn–Teller bipolaron mechanism of high-T_c superconductivity that Bednorz and Müller were hinting at. I sat down and developed a multi-band model, incorporating Jahn–Teller effect.

A week later I reported my progress to Anderson. He listened to me and made a brief remark, "Your theory is beautiful, but is totally irrelevant for cuprate superconductivity." My level of excitement about my theory was high. I was deaf to Anderson's critical remark. By then I knew that the parent compound La_2CuO_4 is a Mott insulator, thanks to some papers on La_2CuO_4 given by T. P. Radhakrishnan, a graduate student at the Department of Chemistry at that time. But I was insensitive to the nearness to a Mott insulator as well, and kept going my way. I even invited my friend and collaborator Dung-Hai Lee from IBM, Yorktown Heights to Princeton to spend a week, to begin a collaboration on this important problem. Incidentally, Dung-Hai and I had great time as collaborators. We used to talk for hours over the telephone. I also spent two memorable summers at IBM Yorktown Heights, visiting Dung-Hai.

5. Mean Field Theory for Spin Liquids, Pseudo Fermi Surface and Superconductivity in Doped Mott Insulator

It was mid December 1986. Anderson was leaving Princeton to attend the valence fluctuation conference at Bangalore. On the day before he was leaving, as a parting note he told me, "Baskaran, the whole thing is a Mott phenomena — think about it." I am used to short but loaded one or two line remarks from Anderson and took them seriously. In fact, I enjoyed thinking about them, like a puzzle. The present statement got registered in my mind and made some subconscious rumblings. However, I continued on the Jahn–Teller path. I was confident that I could convince Anderson about Jahn–Teller mechanism for cuprates and eagerly awaited his return to the US in January 1987. He returned to Princeton, via Caltech.

I called Anderson the day he arrived at Caltech, to discuss my progress. Phil was quick to divert my attention. He said, "Baskaran, I have seen the light." He further added, *Our starting point is a Mott insulator in a resonating valence bond state. We are sitting on a cusp. Doping produces superconductivity.* These three sentences did magic to me. I told him "I think I understand what you are saying. Will call you back shortly."

At that time Dung-Hai Lee and Zhou Zou were in my office. I elaborated to them my interpretation of Anderson's three sentences. After half an hour I called Anderson and explained to him my understanding of his three sentences. He remarked happily, "you have smelled it right." That was the beginning of an enjoyable collaboration on RVB theory high-T_c superconductivity with Anderson. In a week's time a manuscript on a mean field theory of quantum spin liquid and RVB mechanism of superconductivity,[5] co-authored with Anderson and Zou, was ready. Zou was a very smart graduate student and a good collaborator.

Another happy coincidence. During my random walks in Anderson's garden in 1986, but a few months before cuprates arrived on the scene, I read Anderson's 1973 RVB theory paper,[4] a key and fundamental paper on quantum spin liquids, twice. I also wondered about the meaning of phase coherence among the valence bonds in an insulating state. I don't remember why I read this particular paper. But this visit to Anderson's garden prepared me to face the occasion.

Our paper with Zou and Anderson[5] was the first to apply mean field theory to undoped and doped Mott insulators, focus on the constituent electron degree of freedom, rather than local moments. It boldly worked on an enlarged Hilbert space and suggested that incorporation of phase fluctuation should not change the physics qualitatively, based on simple estimates. An idea to decouple the spin-spin interaction term in terms of Cooper pair operator, came from a paper of Noga[30] written in the context of Anderson lattice. A pseudo Fermi surface and a quantum spin liquid in the Mott insulator was gotten with practically no efforts. Soon slave particle methods and Gutzwiller approximation scheme, technically more suited methods followed.

How our article got published is interesting in itself. If Anderson had a fear for anyone in the field of physics, it was the journal referees. I can very well imagine fear on the other side: ending up as a referee for Anderson's paper! Anderson corrected our manuscript and I submitted it to PRL, a natural destination. On knowing this Anderson was worried. He suggested withdrawal from PRL and re-submission to *Solid State Communications*. His concern was a potential delay from referees, because ideas were new and somewhat radical. The irony was, in that exciting initial months following high-T_c cuprate discovery, papers on high-T_c cuprates received by PRL were apparently refereed by a panel of experts, which made instant decisions. PRL had just accepted our paper, when our withdrawal request reached them. We were unaware of it. Respecting our withdrawal request PRL obliged. It is *Solid State Communications* now. After some exchange with a referee and consequent delay, it got published.

6. Emergent Gauge Fields

At IISc, Bangalore, I had a very good course on field theory by R. Rajaraman. He was an outstanding teacher, like his youngest brother R. Shankar at Yale. Rajaraman made us feel at home with difficult subjects. Soon after this exposure, I listened to a lecture series on lattice gauge theory and renormalization group by Leo Kadanoff, another series by Franz Wegner at a Summer School on Statistical Mechanics (1976) at Sitges, Spain. This got me interested in non-abelian lattice gauge theory, strong coupling approach to glue ball mass, etc.

During 1980, while working at the Department of Theoretical Physics of the University of Madras, I gave a short course on lattice gauge theory, to students and colleagues who were all high energy physicists. Field theory on a lattice made things more transparent to me. Elitzur's theorem, on impossibility of spontaneous breaking of a local gauge symmetry became easy. Being in a group of very active high energy physicists for nearly four years, I got exposed to a variety of challenging high energy physics problems and quantum field theory issues.

The projective aspect, namely Hilbert space restriction resulting from removal of double occupancy (Gutzwiller projection) in the low energy description of Mott insulator and doped Mott insulator is of paramount importance in Anderson's starting point. It became clear to me that this Hilbert space restriction in the Mott insulator produces an emergent local $U(1)$ symmetry, and dynamically generated gauge fields, when we describe physics in terms of the underlying physical electrons. I used Elitzur's theorem, a consequence of emergent local $U(1)$ symmetry, to prove that ODLRO exhibited by our mean field is only artificial and it can be easily removed. The phase fluctuations, not manifest in the spin language, captures spin singlet and spin dynamics, leading to dynamical gauge fields. Doping converts the local $U(1)$ symmetry to a global one and allows for possibility of superconductivity and violation of Elitzuer theorem.

I communicated the calculations to Anderson around March 1987. Phil and his wife Joyce were spending their usual spring break, a month of retreat at Cornwall, a coastal village in UK. Anderson replied that he has come to a similar conclusion and explained it. This resonance and a chance for another collaboration got me even more excited. This is the origin of our paper[6] on "Gauge Theory of High-Temperature Superconductors and Strongly Correlated Fermi Systems."

Our free energy for the spin liquid state in a Mott insulator, in terms of the bond singlet variable, has a local $U(1)$ symmetry. Doping reduces the symmetry to a global $U(1)$ symmetry. What is important is that our action, a generalized Ginzburg–Landau free energy expression for doped Mott insulators, had memory of the Mott insulator in the superconducting state. Interestingly, our free energy also had a d-wave solution as the lowest energy solution. However, we were sticking to extended-s symmetry solution because of the then available experimental results, that Anderson respected immensely. The feeling was that strong $U(1)$ gauge field fluctuations will stabilize an extended-s wave solution eventually. This turned out to be not the case. The d-wave solution that Kotlier–Liu, Michael Ma and others found has stood the experimental test.

In our paper we hinted at a hidden particle-hole symmetry in addition to the local $U(1)$ symmetry. The two got nicely woven into a beautiful $SU(2)$ local gauge theory by Anderson, Ian Affleck, Zhou Zou and Ted Hsu. Gauge field as an emergent low energy degree of freedom was welcome by the community. Very soon connection of the magnetic fluxes and electric fields of the emergent $U(1)$ fields to spin current (chirality) and valence bond dynamics was established.

7. Spin Charge Decoupling, Anomalous Metallic State, etc.

As we were understanding mechanism of superconductivity and developing approximation methods to study superconducting state, it became clear that Anderson's mind was getting diverted into the anomalous normal state of the optimally doped cuprate. Interestingly, Anderson was very satisfied with local superexchange as the pairing mechanism and origin of a strong pairing scale, our RVB mean field theory and several important notions and scenarios that emerged with it. The famous temperature-doping phase diagram with a dome, was predicted on very general grounds in our PRL paper in 1987,[7] even before such a phase diagram was experimentally measured. While Kivelson, Rokhsar and Sethna[16] coined the name holon, the unpaired neutral fermion that Anderson introduced in his 1987 paper remained nameless. Anderson christened it "spinon," in our paper.

The prediction of spin charge decoupling by Kivelson, Rokhsar and Sethna was brought into somewhat sharp focus in the optimally doped metallic state in this paper. This paper also talked about possible linear resistivity in the normal state, based on a simple golden rule estimation of scattering of (incoherent) holons by the fermionic spinon quasiparticle excitations from the pseudo Fermi surface. It emphasized that the doping of the Mott insulator does not produce a rigid displacement of underlying (spinon) Fermi surface.

While Anderson respected that a theory should be able to describe physics qualitatively and quantitatively, he also realized that such a theory is going to be tough, because of the projection and a consequent strong coupling character; one should not get intimidated by difficulties and ignore deep insights that one can get from mean field type of theories and physics motivated approximate theories. Similarly model building is an important part of the game. A model should be simple and not simpler (sic).

It also became clear to us that t-J model, introduced by Anderson to describe cuprate physics is more appropriate for the optimally doped region. Underdoped, pseudo gap region is dominated by residual unscreened Coulomb interactions, disorder effects and electron-phonon coupling, not contained in the t-J model. The best way to understand the mechanism of superconductivity in its purest form is the optimally doped regime. Anderson used to call experimental phenomenon seen in under doped regime as nanoscopic phase separation and metallurgical complications.

However, thanks to continuing experimental efforts, new physics such as existence of "small Fermi suface," hiding in the background of a variety of competing phases, has emerged from quantum oscillation experiments, for example. This is unexpected, as we have a strong memory of the Mott insulator in the form of very ill defined quasiparticles at the chemical potential as shown by ARPES.

I have offered an explanation[8] for existence of a pseudo Fermi liquid having a small Fermi surface by applying the idea of Haldane exclusion statistics to the doped hole in a short range spin liquid in a Mott insulator. That is, doped hole is a composite object, a loosely bound (by RVB gauge forces) topological excitations, namely holon and a spinon, in a reference neutral spin liquid state, rather than a band insulator. Holon and spinon have Haldane exclusion statistics of 1 and $\frac{1}{2}$, and gives doped hole an exclusion statistics of $\frac{3}{2}$. I have called this Fermi liquid-like state a $\frac{3}{2}$ Fermi liquid. An ordinary fermionic hole in a band insulator for example, has an exclusion statistic of 1 and occupies one unit of volume in k-space. However, each $3/2$ fermion occupies, on average, $3/2$ times the volume k-space. It results in an expanded Fermi pocket radius, consistent with $\sim 3/2$ times expansion seen in experiments.

8. Are We Bees?

I had a wonderful visit to Aspen in the summer of 1987 (see Fig. 2). By that time I was convinced that the resonating valence bond states advocated by Pauling in the context of p-π bonded molecules and graphite, and elevated to novel quantum spin liquid states in the Mott insulator in 1973 by Anderson should be ubiquitous beyond cuprates. Anderson and I had talked about possible role of RVB physics, in doped $BaBiO_3$ and so called *bad actors* A15 and Chevral phase superconductors. $BaBiO_3$ is interesting. It was popularly known as a negative U Hubbard system because of valence skipping and charge disproportionation. However, an early analysis of spectroscopic results by Kasuya, did not support the charge disproportionation and negative U idea. This gave Anderson and me confidence to think about repulsive Hubbard model for doped $BaBiO_3$.

This thought, that there may be other systems with RVB physics was high in my mind at Aspen. Aspen is a beautiful mountain resort and a great place to contemplate. There is music in the background, flowers, mountains and possibilities of nice hikes in the summer; small number of lectures and

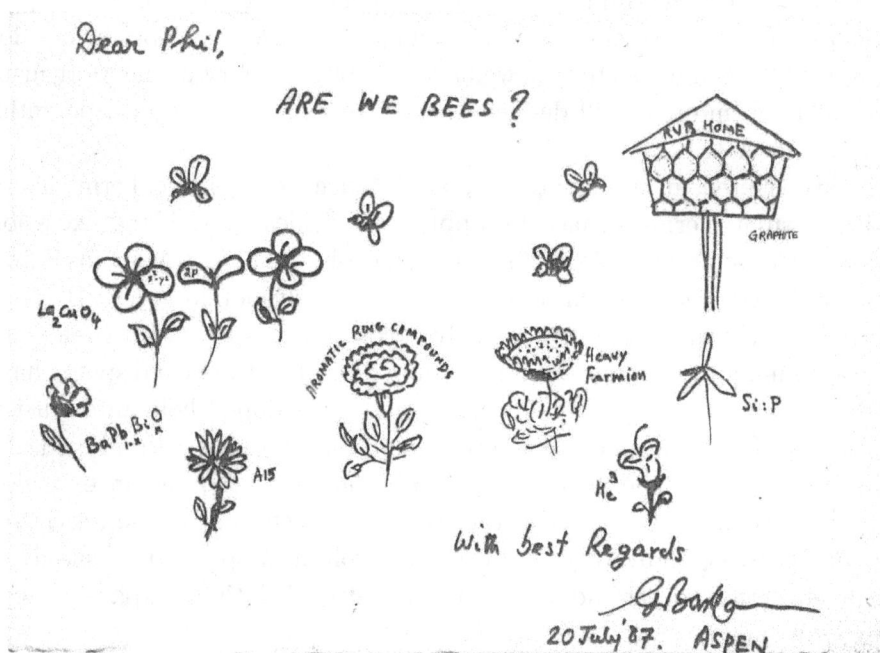

Fig. 2.   *Are We Bees?* A hand-drawn picture postcard by the author to Anderson in 1987 from Aspen. In addition to cuprates, it also depicts potential new materials where valence bond resonance and interesting physics, including superconductivity might be residing.

generally free, welcome and friendly atmosphere. Among several others, people like Anderson, Elihu Abrahams and Ravin Bhatt had worked hard to maintain this as a great Physics Center in the world.

Having been obsessed with cuprates, flower petals in the garden reminded me of d-orbitals of copper and $2p$ orbitals of oxygen in the CuO_2 layers. I took that opportunity to make a sketch and posted it to Anderson (Fig. 2). Anderson liked the picture and put it as a slide in one of his talks. This picture appears in an AIP conference proceedings, with Anderson's comments.[31]

Looking back, this picture emphasized graphite (a single layer of which is graphene) as RVB home. In the wake of superconductivity in MgB_2 this idea surfaced again[9] and the result was my prediction of very high T_c in doped single sheet graphite, based on a t-J-0 model, that is, a t-J model with no double occupancy constraint. Graphite is not a Mott insulator; it is a highly anisotropic semi-metal. However, it has a strong nearest neighbor covalent bond or singlet pair correlation, according to Pauling. The t-J-0 model I introduced thus combined band physics with valence bond (spin singlet physics) in a semi-microscopic phenomenological fashion. To my great satisfaction, my work was pursued by good friend Seb Doniach and his student Annica Black-Schaffer.[10] They discovered an unconventional $d + id$ chiral superconductivity, with a very high superconducting T_c. A variational Monte Carlo calculation[11] with Vijay Shenoy and Sandeep Pathak, that took into account quantum fluctuations beyond mean field theory brought the scale of T_c down. It was still high and a welcome value of 200K! Experiments have not confirmed our prediction, possibly because of an unavoidable disorder that comes at the desired range of doping and a high sensitivity of $d + id$ state to disorder. Such a possibility in silicene and germanene, as I have suggested[12] makes it even more exciting.

This picture has P-doped Si, with four flower petals mimicking the sp^3 tetrahedron bonds. I had imagined a correlation based narrow impurity bond superconductivity. Later experiments showed superconductivity, but the scale was too low, below 1K. Fortunately B-doped diamond exhibited a higher T_c of 12K later in 2004. I had attempted to explain this[13] as a self-doped impurity band Mott insulator superconductivity, occurring close to the Anderson–Mott transition point, in the doping axis.

I don't dare say I predicted superconducting C_{60} compound, even though the big flower (with "aromatic ring compounds" written on top) was supposed to mimic a big molecule with p-π bonds and ring currents. $BaBiO_3$ and A15 compounds that keep coming to my mind once in a while even now are there in the picture.

The message of this section is that *Anderson inspires*.

9. Return to India. Beyond Cuprates and a Synthesis

Having accepted a job at the Institute of Mathematical Sciences, Chennai, I decided to return to India in late 1987. It was a difficult decision, as my collaboration with Anderson with high-T_c superconductivity made me popular, with tempting offers from some outstanding places in the US and elsewhere. In a span of seven months in 1987, after I began my work with Anderson on high-T_c superconductivity, I gave nearly 35 talks all over north America, Japan and a few places in Europe. As there were new concepts, notions and techniques, my talks often got stretched to several hours of discussions and working out details. The maximum was, 12 hours, at Hide Fukuyama's group in Tokyo in the summer of 1987. It was similar when I visited Vinay Ambegaokar at Cornell. I was excited by the captive audience, wherever I went. Probably the audience got enthused by my own excitement.

Soon I became an *RVB preacher*. ICTP, Trieste became a good ground for this. I got involved in various activities on strongly correlated electron systems and high-T_c superconductivity on a regular basis, thanks to invitation and support from Erio Tosatti, Yu Lu, Mario Tosi, Stig Lundqvist and Abdus Salam. I have been regularly visiting Princeton, almost on yearly basis and continuing my collaboration with Phil Anderson. One such visit was nearly two years stay (1995–1996) at the Institute for Advanced Study, Princeton and the Physics Department of Princeton University.

According to my friends I always smelled RVB theory in any new superconductors that emerged on the scene. It started with, fullerites like K_3C_{60}. With Erio Tosatti we developed a mechanism for superconductivity[14] that used the valence bond correlations in the fullerene molecules and led to pair binding, a notion that was independently introduced by Kivelson and Chakrabarty.[15] Anderson was very supportive of our theory. Molecular conduction bands in K_3C_{60} were very narrow, less than 0.25 eV. To Anderson it was a surprise that the stoichiometric compound K_3C_{60} manages to be a metal. Mott localization should be imminent. Later experiments by Iwasa and others showed that indeed K_3C_{60} can be converted into a Mott insulator by a negative pressure caused by an expansion, just by inserting certain number of inert NH_3 molecules into interstitial space in K_3C_{60} in the unit cell.

The next superconductor in line was the nickel borocarbide family, discovered by the TIFR group at Mumbai and Cava and others in the US. It was a layered system. Interestingly the layer structure is similar to the FeAs layers in the Fe-based superconductors. I was convinced of a mechanism of superconductivity based on electron correlation.[18] Neutron scattering indicated a strong (π, π) type magnetic fluctuations, and even a signal for a neutron resonance mode. In view of a complicated-looking band structure, modeling was somewhat complicated and a deep understanding remains obscure.

To me organic superconductors, often in the vicinity of a Mott insulator was a mystery for a long time. How does pressure convert a Mott insulator into a supercondctor? I realized, from the existing pheomenology that the superconducting side of Mott transition point is better thought of as a self-doped Mott insulator. The self-generated holons and doublons, of equal density, in the half-filled band is determined by some kind of Madelung energy again, in line with an inevitable long range interaction on the Mott insulating side; but Hubbard model misses this. So I suggested a two species t-J model to understand superconductivity in organics.[17] This unified superconductvity in cuprates and organics.

Then came $Na_x CoO_2 \cdot yH_2O$, a hydrated sodium cobalt oxide superconductor. Narrow band and correlation based physics superconductivity was obvious. I predicted a $d + id$ RVB type of superconductivity.[19] Charge ordering in the CO_2 layer and ordering in the intercalant Na layer and role of H_2O molecule complicated the physics and being able to understand the experiments in detail.

Fe arsenide superconductor seems to be a good example of double RVB system,[20] where two valence electrons in the $3d^6$ shell of Fe^{2+} form some kind of Hund coupled $2d$ spin-$\frac{1}{2}$ RVB system with internal charge transfer. Unfortunately this family is also complex, unlike the cuprates, where a single band makes life much simpler. The cousin of Fe arsenide, FeSe and FeS seem to be making things simpler. A better understanding of superconductivity might emerge from these systems.

Silicene, germanene and stanene are $2d$ analogues of graphene. C, Si, Ge and Sn occur in the same column in the periodic table. However, the atomic radii of Si and Ge are about 60% larger than that of carbon. According to electronic structure calculations this leads to a substantial, three-fold reduction in the bandwidth of p-π band in silicene and germanene. Based on band theory estimates of the t and U parameter and some overwhelming phenomenology I came to the conclusion that silicene and germanene are likely to be Mott insulators.[12] This is a prediction that is yet to be confirmed, because of not being able to synthesize free-standing silicene or silicene on insulating substrates. The t and J parameter I estimate for silicene makes it a prospective playground for room temperature superconductor, provided competing phases such as valence bond ordering can be kept under control.

Few other interesting systems of our interest are superconductivity in spin ladder compound[22] and a recently popular doped $TiSe_2$[21] with potential chiral spin singlet superconductivity, in line with earlier doped graphene, hydrated cobalt oxide.

Thanks to Piers Coleman's comments and criticism at ICTP Trieste, I ended up predicting[23] p-wave superconductivity in Sr_2RuO_4, independently of Rice and Sigrist. Interestingly, Infinite U Hubbard model, according to our recent study[24] also supports chiral $p + ip$ superconductibity riding on top of a Nagaoka ferromagnetism.

Based on my experience with cuprates, thanks to Anderson and later works indicated above, I suggested a *Five-fold Way to New Superconductors*.[25] Here electron correlation and relatively narrow bands play central role. The five routes are: (1) Copper Route (doped spin-$\frac{1}{2}$ Mott insulator), (2) Pressure Route (Self-doped Mott insulator in organics), (3) Diamond Route (RVB physics in impurity band and superconductivity), (4) Graphene Route (Broadband $2d$ and intermediate correlations) and (5) Double RVB Route (Fe arsenide and self-doped spin-1 Mott insulator).

10. Theory of Cooper Pair Crystals and Superconductivity in Molecular Crystals Under Pressure

Solid hydrogen has evoked a great interest over decades, from condensed matter to planetary physics community. Wigner and Huntington predicted in 1935[26] that solid H_2 will metallize at a pressure ~25 GPa. Ashcroft predicted in 1968[27] that such a metallic hydrogen will be a room temperature BCS superconductor. Unfortunately both predictions have not been confirmed experimentally. Solid H_2 refuses to metallize even at a few hundred GPa. Instead, it undergoes a series of structural changes, where covalent bond reorganization takes place; it remains insulating. In view of this, interest in the community shifted to hydrogen rich solids. Silane, SiH_4, has yielded under pressure and becomes superconducting with a modest $T_c \sim 20K$, as shown by Eremets *et al.*[28] Interestingly, in a recent paper Eremets *et al.*, also report superconductivity[3] in another hydrogen rich solid, H_2S, and a much higher $T_c \sim 203K$ at a pressure of 200 GPa.

These exciting results need to be reconfirmed by measurement of Meissner effect, Josephson tunneling, etc. A few theory papers have appeared before and after the experiment, based on *ab initio* calculation. They predict a variety of structures, and a high-T_c superconductivity based on electron-phonon interaction mechanism. Unfortunately, the structure of H_2S is not known experimentally. In an earlier (2004) experiment X-ray scattering revealed a structure at high pressures. There is molecular dissociation and formation of sulfur-sulfur bond and helical strings, known in allotropes of S, Se and Te. Position of hydrogen atom could not be determined, in view of a low electron number associated with H atom. In the absence of phase separation, H atoms or H molecules are likely to occupy small interstitial positions in the densely packed sulfur chain lattice. Sulfur atom has a large atomic radius compared to H atom.

In our theory[29] we suggest an organization principle, where covalent bonds continue to survive, but may change their spatial pattern. This arises from the kinetic energy gain in every covalent bond by the electron pair being in an orbitally symmetric state. That is, two electrons behave like bosons occupying the same quantum state (albeit with some correlation hole). Antisymmetry is taken care of by the singlet spin state. From phenomenology and from study of available structures in theory and experiments in H_2 and H_2S, I find a strong tendency for covalent bond conservation. There is a resistance to form simple unfilled bands and Fermi sea.

I view these paired valence electrons that form covalent bonds in molecules as stable but confined Cooper pairs. Thus in H_2S the two saturated covalent bonds correspond to two confined Cooper pairs. Molecular solid H_2 and H_2S are in this sense Cooper pair insulators. They are so deeply bound that this is irrelevant in normal situations. However, high pressure, a peculiar liberator, could deconfine some of these confined Cooper pairs and create a superconducting state. That is pressure, under some conditions and for a range or pressures, could liberate electron pairs and not single electrons.

Specifically I suggest the following structural and valence bond reorganization for the superconducting H_2S at high pressures. It leads to strong sulfur-sulfur bond and helical chains or a structure with a dominant saturated S-S bonds. Some dissociated H atoms can remain neutral H-atoms and some as H_2 molecules. At high pressure with a strong S-S bonds, S atoms form a dense packing. Because of its large atomic radius S atoms leave only small interstitial space for H atom or H_2 molecule. The interstitial crystalline network which accommodates H atoms, can be quasi-1-, quasi-2-dimensional or 3-dimensional. The H-H distance and H sublattice structure is now dictated by the sulfur sublattice. In general the structure does not allow for saturated H-H bonds. For example a dimerized H-atom chain allows for saturated H-H bonds. Whereas, uniform chain leads to valence bond resonance. I find that in some pressure ranges, the H-H transfer matrix elements (direct and through sulfur atoms) is comparable in size to the ionization energy of H atom, leading to the possibility of a Mott insulating neutral H atoms sublattice. As mentioned earlier, in structures like uniform linear chain, or hexagonal sheet or fcc lattice, valence bonds are not saturated in general. This leads to a possibility of quantum mechanical resonance of valence bonds.

In my picture, pressure and structural changes, under some conditions, help form a Mott insulating sublattice of neutral H atoms. This state is likely to be an unstable insulating state. Because, in general, a differing electronegativity of the sulfur and H sublattices will result in a charge transfer between the two subsystems. This charge transfer dopes the H atom Mott insulator and opens the door for superconductivity in a doped Mott insulator. My estimates of the doped Mott insulator parameters, for a few structures for H_2S available in the literature from LDA calculation gives the possibility of superconductivity reaching the scale of 200K, as seen in the experiment. Pressure induced dissociation in H_2S has been suggested to create a new compound H_3S, following a phase separation. Our picture goes through for hydrogen rich solids such as H_3S and a recent pressure induced superconductor[32] P_3S, with a $T_c \sim K$.

Acknowledgments

Interaction and collaboration with Anderson, since 1983 has been inspiring and wonderful. It is indeed amazing that one man could do so much to science. It is equally amazing that one man could influence so much, those who cross his path. It gives me great pleasure to congratulate Anderson on this happy occasion and wish him, Joyce and Susan all well. I thank the Science and Engineering Research Board (SERB), India for the SERB Distinguished Fellowship. This research was supported by Perimeter Institute for Theoretical Physics, Waterloo, Ontario, Canada.

References

1. A. Bednorz and A. Müller, *Z. Phys. B* **64**, 189 (1986).
2. P.W. Anderson, *Science*, **235**, 1196 (1987).
3. A.P. Drozdov *et al.*, Nature, **doi: 10.1038/nature14964** (arXiv:1506.08190) and Mari Einaga *et al.*, arXiv:1509.03156
4. P.W. Anderson, *Mater. Res. Bull.* **8**, 153 (1973).
5. G. Baskaran and P.W. Anderson, *Solid State Commun.* **63**, 973 (1987).
6. G. Baskaran and P.W. Anderson, *Phys. Rev. B* **37**, 580 (1988).
7. P.W. Anderson, G. Baskaran, Z. Zou and T. Hsu, *Phys. Rev. Lett.* **58**, 2790 (1987).

8. G. Baskaran, arXiv:0709.0902
9. G. Baskaran, *Phys. Rev. B* **65**, 212403 (2002).
10. A.M. Black-Schaffer and S. Doniach, *Phys. Rev. B* **75**, 134512 (2007).
11. S. Pathak, V. Shenoy and G. Baskaran, *Phys. Rev. B* **81**, 085431 (2010).
12. G. Baskaran, arXiv:1309.2242
13. G. Baskaran, *J. Supercond. Novel Magnetism* **21**, 45 (2008).
14. G. Baskaran and E. Tosatti, *Current Science* **61**, 33 (1991).
15. S. Chakravarty and S. Kivelson, *Europhys. Lett.* **16**, 751 (1991).
16. S.A. Kivelson, D. Rokhsar and J. Sethna, *Phys. Rev. B* **38**, 8865 (1987).
17. G. Baskaran, *Phys. Rev. Lett.* **90**, 197007 (2003).
18. G. Baskaran, *J. Phys. Chem. Solids* **56**, 1957 (1995).
19. G. Baskaran, *Phys. Rev. Lett.* **91**, 097003 (2003).
20. G. Baskaran, *J. Phys. Soc. Jpn.* **77**, 113713 (2008).
21. R. Ganesh, G. Baskaran, Jeroen van den Brink and Dmitry V. Efremov, *Phys. Rev. Lett.* **113**, 177001 (2014).
22. J.P.L. Faye, S.R. Hassan, P.V. Sriluckshmy, G. Baskaran and D. Sénéchal, *Phys. Rev. B* **91**, 195126 (2015).
23. G. Baskaran, *Physica B* **223 & 224**, 490 (1996).
24. Zheng-Cheng Gu, Hong-Chen Jiang and G. Baskaran, arXiv:1408.6820.
25. G. Baskaran, *Pramana* **73**, 61 (2009).
26. E. Wigner and E. Huntington, *J. Chem. Phys.* **3**, 764 (1935).
27. N.W. Ashcroft, *Phys. Rev. Lett.* **21**, 1748 (1968).
28. M.I. Eremets, I.A. Trojan, S.A. Medvedev, J.S. Tse and Y. Yao, *Science* **319**, 1506 (2008).
29. G. Baskaran, arXiv:1507.03921
30. M. Noga, Czech, *J. Phys.* **38** 210 (1988).
31. P.W. Anderson, in *Modern Physics in America — A Michelson-Morley Centennial Symposium*, Eds. W. Fickinger and K.L. Kowalski, AIP Conference Proceedings **169** (American Institute of Physics, 1988).
32. A.P. Drozdov, M.I. Eremets and I.A. Troyan, arXiv:1508.06224

Some Reminiscences on Anderson Localization

Elihu Abrahams

Department of Physics and Astronomy
University of California, Los Angeles
Los Angeles, CA 90095-1547, USA
abrahams@physics.ucla.edu

On the occasion of PWA's 90th birthday celebration, I review some happenings that contributed to the development of theory and experiment on Anderson Localization, one of the many subjects originated and developed by Phil, which became a central theme in condensed matter physics.

1. Introduction

In this essay, I look back at events connected with the development of the subject of Anderson Localization and in particular, the properties of (mostly) non-interacting degenerate electrons undergoing transport in the presence of disorder. Thus the title "Reminiscences . . .". I apologize to Phillip for looking backward, as I know he would much rather hear about new physics.

2. Before 1958

George Feher was a key player in the story. He and I were friends and graduate students together at Berkeley in the early 1950s. He joined Bell Laboratories after getting his PhD and he facilitated my becoming a summer visitor there when I moved to Rutgers University in the summer of 1956. As it happened, the Bell Labs Theory Department started in that year. In the summers of 1956 and 1957, Bell had a large visitor program that involved theorists of various persuasions and so this is where I first interacted with PWA. I had met him a year or two earlier when I was a graduate student on the occasion of a visit of his to Berkeley. My advisor was Charlie Kittel, who was very opinionated about both physics and people. Consequently, he was at pains to let his students know whom they should like and whom they should not like. Phil was to be liked and and especially to be listened to.

However, it was not until that 1956 summer that I learned how to listen to Phil and to make a start on figuring out what he was saying. It was only later that I learned (from my children) to ask "how do you know that?"

The real reason I mention George Feher is to recall how exquisite experimental advances can stimulate fundamental new theory. The case here concerns George's beautiful electron-spin resonance experiments[1] on phosphorous-doped silicon and his invention of the electron-nuclear double resonance technique (ENDOR). This work led to a number of interesting theoretical papers, but in the present context, I want to emphasize that Phil's famous and crucial 1958 paper[2] was stimulated in part, as he has often said, by George's result on the lack of spin diffusion in phosphorous-doped silicon.

Phil spent a lot of time talking to me that summer about "the absence of diffusion in certain random lattices." His patience and generosity were remarkable and I like to think that my obtuseness was responsible for the opening sentence of Section II of that seminal paper:

> "Since the mathematical development is fairly complicated and involves lengthy consideration of each of a number of points, we should like to summarize the reasoning rather fully in this section, leaving the proofs and details to later sections."

For what it may be worth, I note that by 1968, that 1958 paper had achieved 30 citations (!). By now, it must be over 5000.

3. 1970s, Gang of Four

In the mid-1970s, I was visiting Princeton several times a month to talk with Phil and colleagues. T. V. Ramakrishnan was in residence as a Visiting Professor in 1978 and Don Licciardello, who had been working with David Thouless, was there as an Assistant Professor.

One of the things the three of us, plus Phil, were discussing was the issue of minimum metallic conductivity in two and three dimensions (2D and 3D). Don was a mighty expert on this question and he had written several relevant papers with David Thouless. In the last of these,[3] they found, by numerical work on large square $L \times L$ lattices, up to $L = 40$, values of the conductance appreciably less than the so-called minimum metallic conductivity. See Fig. 1. They said: "It could be argued that σ always scales to zero for 2D systems." So Don already knew that it was likely that in disordered 2D,

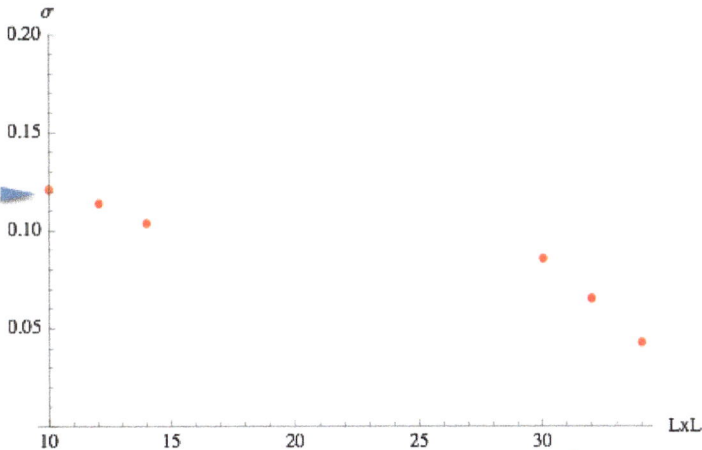

Fig. 1. Numerical data of Licciardello and Thouless. The blue arrow indicates the "minimum metallic conductivity."

there would be no extended states. Nevertheless, absent an analytic solution for the issue, the rest of us were unconvinced.

It happened that in 1978, I noticed a paper of H. G. Schuster in *Zeitschrift für Physik*.[4] He tried to map the behavior of disordered non-interacting electrons in two dimensions onto the 2D XY model. The known jump in the spin stiffness at the transition in that model was related to the value of the conductivity at a mobility edge (the energy that separates extended from localized states) and this produced a minimum metallic conductivity in essential agreement with that found in some earlier simulations by Licciardello and Thouless and with the Ioffe–Regel criterion (the mean free path cannot be less than the quasiparticle wavelength) used by Mott, who had been emphasizing the role of the minimum metallic conductivity.

As mentioned above, Don already suspected that this could not be right, but the Schuster paper got us thinking about how to formulate a scaling description and it directed our attention to the influential work of Franz Wegner,[5] who proposed some scaling laws for conductance near a mobility edge. As it happened, David Thouless had already suggested some years earlier how to set up scaling[6] and we followed his ideas rather closely. The key point was to recognize that the dimensionless conductance was a function of length scale and should be considered as a scaling variable.

Learning from Thouless' approach, we proceeded to formulate a phenomenological discussion, which led to the "Gang of Four" Physical Review Letter,[7] whose noteworthy conclusions included the statement that there could be no extended states in dimension two or less ("absence of diffusion").

Thus the title echoed that of Phil's 1958 paper,[2] although the latter was only referred to in a minor way. Here are the steps of the argument:

(1) Transport is determined by the behavior of the electron wave functions at the Fermi level.
(2) Examining the behavior of the wave functions in the presence of disorder and their sensitivity to boundary conditions leads to consideration of the conductance $G(L)$ of a block of size L in d spatial dimensions.
(3) The dimensionless conductance $g(L) = G/(e^2/\hbar)$ depends on the block size such that $g(L + \delta L)$ depends only on $g(L)$, which thus serves as a scaling variable.
(4) Borrowing from the language of critical phenomena, we examined the "β-function" $\beta(g) \equiv d \ln g/d \ln L$.
(5) What is known: At large g, metallic behavior, so $g = L^{d-2}\sigma \to \beta \sim d-2$. At small g, localized behavior, so $g \propto e^{-L/\xi} \to \beta \sim \ln g$.

By smoothly connecting these asymptotic behaviors of the conductance g, we proposed the behavior of the β-function as seen in Fig. 2.

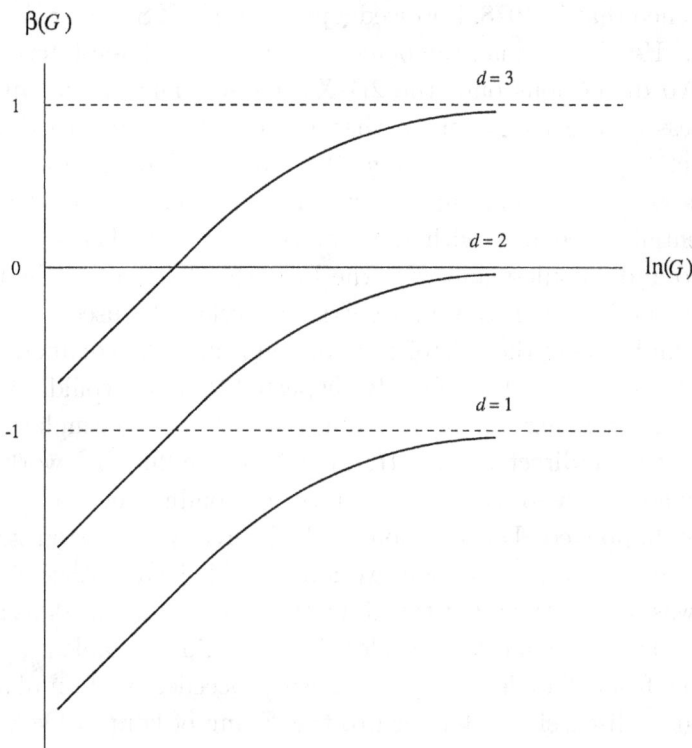

Fig. 2.   "Gang of Four" β-function. G = conductance, d = dimensionsality. L = sample size.

We drew significant conclusions from the behavior shown in the figure: Most important is that only for $d > 2$ does the β-function have a zero, i.e. an unstable fixed point; it corresponds to the existence of a *continuous* metal-insulator transition with no "minimum" metallic conductivity. Correspondingly, at $d = 2$ there should be no conducting (extended) states at all: at sufficiently large L, all states will exhibit exponential decay at the sample boundaries. We also were able to confirm the earlier (1976) conjecture[5] of Franz Wegner that for $d > 2$, the conductivity σ behaves at the transition as $\sigma \propto (x - x_c)^{(d-2)\nu}$, where x_c is the value at the transition of the relevant tuning parameter x, which could be, for example, the quasiparticle energy (in which case x_c would be the energy at the mobility edge). Here, ν is related to the slope of the β-function at its zero.

During our discussions in Princeton, Rama (T. V. Ramakrishnan) found the important interesting paper[8] of Jim Langer and T. Neal, which noticed a breakdown in the calculation of the impurity resistivity as an expansion in powers of impurity concentration. It was due to particular impurity scattering contributions, which when we summed them up, gave a diffusion-like pole in a backscattering channel. This produced a first perturbative correction, for large conductance and it confirmed the leading $1/g$ contribution that we had already conjectured for the β-function: $\beta(g) \sim d - 2 - a/g$.

One of the many "tyrannies" that Phil has often complained about over the years is the tyranny of Feynman graphs. However it must be said that studying the so-called "crossed graphs" of Langer and Neal became the standard approach to the problem of the Anderson localization of non-interacting electrons in a random potential. An example is shown in Fig. 3, where it may be seen that the dotted impurity scattering lines are maximally crossed if one unfolds the diagram.

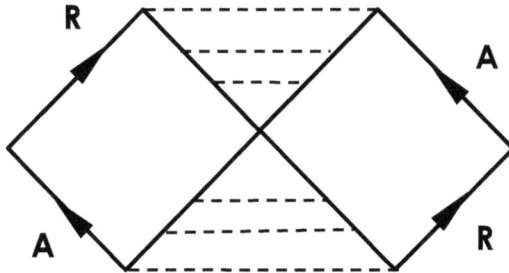

Fig. 3. Feynman diagram for impurity-averaged conductivity. R and A denote retarded and advanced Green's functions. The dashed lines represent impurity scatterings after averaging.

4. Anderson Localization Post-Gang of Four

Immediately after the Gang of Four paper appeared, Shinobu Hikami, Tolya Larkin, and Yosuke Nagaoka[9] ("HLN") followed up and included spin-orbit scattering, spin-flip scattering and magnetic field into the maximally crossed graphs analysis. Their beautiful formula is still used to analyze experimental results on two-dimensional disordered metals. At the same time, various experimental results appeared and they were in support of the theory. The scientific atmosphere at Bell Laboratories, indeed its very existence, made such rapid and exciting advances possible. In particular, Gerry Dolan and Doug Osheroff found $\log T$ temperature dependence of resistance in Au-Pd thin films[10] and David Bishop, Dan Tsui and Bob Dynes found the same in Si-MOSFET.[11] The latter system, because of the ability to control the charge carrier concentration, became the popular choice for later studies of transport in two dimensions.

The $\log T$ arises because the theory contains a small momentum cutoff, originally $1/L$. It was natural for us, again following Thouless, to argue that a natural replacement for L is the "Thouless length, that is, the diffusion distance in a temperature-dependent inelastic scattering time. This gives a $\log T$ dependence,[12] as observed.

On the subject of weak localization and experiment, I want to mention the excellent experimental results and the review written by Gerd Bergmann in 1984 *Physics Reports*.[13] Here, he introduced the physically appealing cartoon of interfering time-reversed paths to explain the coherent backscattering enhancement that underlies the theory. His results for the magneto-resistance of thin metal films reproduced precisely the results of Hikami, Larkin and Nagaoka already mentioned and clearly demonstrated the weak anti-localization effect that arises from spin-orbit scattering.

5. Electron-Electron Interaction

Naturally, the effect of electron-electron interaction on these considerations became of immediate interest. At about the same time, in fact somewhat before, it was recognized that impurity scattering and the consequent diffusive motion of electrons would influence the interaction and its affect on transport and thermodynamic properties, particularly in two dimensions. Boris Altshuler and Arkady Aronov did pioneering research[14] on the latter issue. They subsequently joined with Patrick Lee in 1980 to produce the definitive work on the perturbative effects of electron-electron interaction in disordered two-dimensional metallic systems, and, it must be said, again making excellent use of the Feynman graph techniques.[15]

Meanwhile, in the mid-1980s, several contributions were made by several people, including Carlo DiCastro, Claudio Castellani, Patrick Lee, Gabi Kotliar, Sasha Finkelstein and Kostya Efetov, developing a renormalization group ("RG") formulation for the the Anderson localization problem in terms of a matrix non-linear σ model as first proposed by Franz Wegner and co-workers in the 1970s. The connections between the perturbative Feynman graph approach and the non-linear σ model RG method were explicated rather neatly by Shinobu Hikami, in 1981.[16] These formal developments were important as they laid out the technique to give a unified RG description, which could treat disorder and interactions on an equal footing, as shown by Finkel'stein.[17]

6. Metallic States in Two Dimensions?

As explained above, one of the striking conclusions of the Gang of Four paper was that a two-dimensional disordered metal would have no conducting states and no metal-insulator transition. Of course, many experiments confirmed this! However, In 1994, things took a very serious turn when Sergey Kravchenko, Volodya Pudalov and co-workers reported the observation of metallic conduction in 2D Silicon MOSFET.[18,19] This violation of the by then well-established canon marked the end of the decade during which the Gang of Four paper was cited enthusiastically at every opportunity.

Some typical data is shown in Fig. 4, in which there appears to be a critical electron density $n_c \simeq 0.9 \times 10^{11}$ cm^{-2} that divides conducting states from insulating ones. Even the most casual examination of the data reveals the strong suggestion of the existence of a quantum critical point that defines a zero-temperature metal-insulator transition. Of course, in spite of Phil's sometime-stated complaints about another tyranny — the tyranny of quantum criticality, it is perfectly natural to expect that a true metal-insulator transition is a quantum phase transition as neither metal nor insulator are clearly defined at non-zero temperature.

In recent years, RG flow diagrams and metal-insulator-Wigner crystal phase diagrams have been produced by various theorists using various techniques — all in efforts to elucidate the physics of disorder plus electron-electron interaction in two dimensions and above. However, there is not yet real agreement as to the final resolution. Some of the non-convergent results

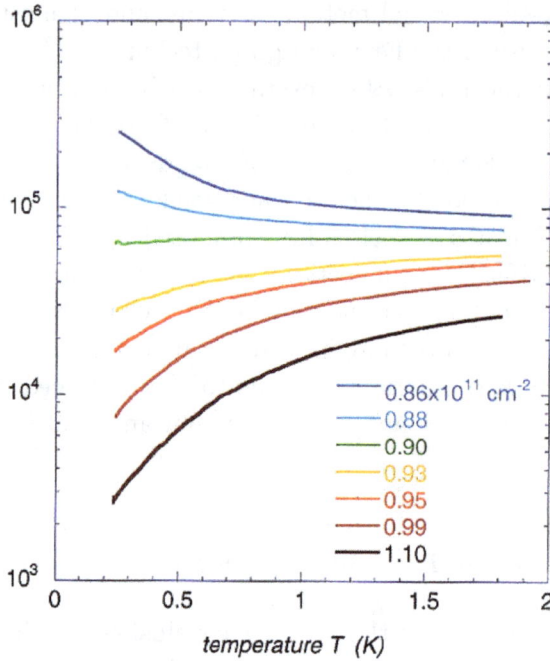

Fig. 4. Resistivity (in ohms) as a function of temperature for the two-dimensional system of electrons in a silicon MOSFET. Different curves are for different electron densities. From Ref. 19.

may be reviewed by consulting several of the contributions in a recent review volume.[20]

This exhausts the years of my reminiscences, so I conclude with some remarks and questions concerning why the physics of Anderson localization remains fascinating and important.

- Realizations of the two-dimensional case are made in high-mobility semi-conductor heterostructures at very low carrier density ($r_s \geq 10$).
- Since the dimensionless interparticle spacing r_s is also the ratio of inter-action energy to kinetic energy, this is in a strong-coupling regime, thus a strongly-correlated electron fluid with disorder.
- Is the metallic phase a "non-Fermi liquid"?
- Is the "metal-insulator transition" governed by a quantum critical point?
- Is Anderson localization robust against interactions?
- What is "many-body localization" and its experimental manifestations?

References

1. G. Feher, Electron spin resonance experiments on donors in silicon. I. Electronic structure of donors by the electron nuclear double resonance technique, *Phys. Rev.* **114**, 1219–1244 (1959).
2. P.W. Anderson, Absence of diffusion in certain random lattices, *Phys. Rev.* **109**, 1492–1505 (1959).
3. D.C. Licciardello and D.J. Thouless, Conductivity and mobility edges in disordered systems II. Further calculations for square and diamond lattices, *J. Phys. C: Solid State Phys. B* **11**, 925–936 (1978).
4. H.G. Schuster, On a relation between the mobility edge problem and an isotropic XY model, *Z. Phys. B* **31**, 99–104 (1978).
5. F.J. Wegner, Electrons in disordered systems. Scaling near the mobility edge, *Z. Phys. B* **25**, 327–337 (1976).
6. D.J. Thouless, Electrons in disordered systems and the theory of localization, *Phys. Rep.* **13C**, 93–142 (1974).
7. E. Abrahams, P.W. Anderson, D.C. Licciardello and T.V. Ramakrishnan, Scaling theory of localization: Absence of quantum diffusion in two dimensions, *Phys. Rev. Lett.* **42**, 673–676 (1979).
8. J.S. Langer and T. Neal, Breakdown of the concentration expansion for the impurity resistivity of metals, *Phys. Rev. Lett.* **22**, 984–986 (1966).
9. S. Hikami, A.I. Larkin and Y. Nagaoka, Spin-orbit interaction and magnetoresistance in the two dimensional random system, *Prog. Theor. Phys.* **63**, 707–710 (1980).
10. G.J. Dolan and D.D. Osheroff, Nonmetallic conduction in thin metal films at low temperatures, *Phys. Rev. Lett.* **43**, 721–724 (1979).
11. D.J. Bishop, D.C. Tsui and R.C. Dynes, Nonmetallic conduction in electron inversion layers at low temperatures, *Phys. Rev. Lett.* **44**, 1153–1156 (1980).
12. P.W. Anderson, E. Abrahams and T.V. Ramakrishnan, Possible explanation of nonlinear conductivity in thin-film metal wires, *Phys. Rev. Lett.* **43**, 718–720 (1979).
13. G. Bergmann, Weak localization in thin films, *Phys. Rep.* **107**, 1–58 (1984).
14. For example, B. L. Altshuler and A. G. Aronov, Zero bias anomaly in tunnel resistance and electron-electron interaction, *Solid State Commun.* **30**, 115–117 (1979).
15. B.L. Altshuler, A.G. Aronov and P.A. Lee, Interaction effects in disordered Fermi systems in two dimensions, *Phys. Rev. Lett.* **44**, 1288–1291 (1980).
16. Shinobu Hikami, Anderson localization in a nonlinear-σ-model representation, *Phys. Rev. B* **24**, 2672–2679 (1981).
17. A.M. Finkel'stein, Weak-localization and Coulomb interactions in disordered systems, *Z. Phys. B: Condens. Matter* **56**, 189–196 (1984).
18. S.V. Kravchenko *et al.*, Possible metal-insulator transition at $B = 0$ in two dimensions, *Phys. Rev. B* **50**, 8039–8042 (1994).
19. M.P. Sarachik and S.V. Kravchenko, Novel phenomena in dilute electron systems in two dimensions, *Rev. Mod. Phys.* **96**, 5900–5902 (1999).
20. *50 Years of Anderson Localization*, E. Abrahams, ed. (World Scientific, Singapore, 2010).

Anderson and Condensed Matter Physics

T. V. Ramakrishnan

Department of Physics, Indian Institute of Science
Bangalore 560 012, India and
Department of Physics, Banaras Hindu University
Varanasi 221 005, India
tvrama2002@yahoo.co.in

The legacy of P. W. Anderson, perhaps the most fertile and influential condensed matter physicist of the second half of the twentieth century, is briefly mentioned here. I note three pervasive values. They are: emergence with its constant tendency to surprise us and to stretch our imagination, the Baconian emphasis on the experimental moorings of modern science, and mechanism as the explanatory core. Out of his work, which is spread over more than six decades and in many ways has charted modern condensed matter physics, nearly a dozen seminal contributions, chosen idiosyncratically, are mentioned at the risk of leaving out many which may also have started subfields. Some of these are: antiferromagnestism and broken symmetry, superexchange and strong electron correlations, localization in disordered systems, gauge invariance and mass, and the resonating valence bond in magnetic systems as well as in high-temperature superconductivity.

This is a unique occasion where a large number of those who have given shape to the enterprise of condensed matter physics have gathered to associate themselves with an iconic physicist, perhaps the most influential in our field in the second half of the twentieth century and the most fertile with new ideas.

I would like to start by mentioning something that stands out after these many years. When I began working with Phil in Princeton, around September 1978 or so, many conversations with him left me completely befuddled. This was hard on a rusty physicist low on self confidence. After a few weeks of this, I felt I must "unburden" myself to J. M. (Quin) Luttinger at Columbia, my PhD mentor. When I talked with him about it, he said "Phil does not get things out beyond a certain point of clarity. But he is the most talented person in our field." (Phil Anderson was a co-author in a

Physics Today obituary for Luttinger which specially noted the clarity of his published work and of his lectures. I myself found that his classroom lectures had a deceptive simplicity.) As months passed, I began to make some sense of his association of ideas and phenomena, and indeed became an interpreter in that transformative period of 1978–1981.

The subfield of physics for which the name condensed matter physics was coined by Anderson and Heine in Cambridge in the sixties started basically as a fringe activity (solid state physics, squalid state physics) in the first half of the twentieth century. This "cloud no longer than a man's hand," grew and grew in its second half to cover about two-thirds of the sky of physics (I think that condensed matter physics is a sunny subject, so the analogy does not carry far). It is a specially daunting task to do justice to the title partly because many of the people who made it possible are here in the meeting which stands for a whole era. I am reminded of a verse of the great Sanskrit poet Kalidasa. In the beginning of one of his epics, he said "where is the lineage of the Sun, and where is my mind which goes to small things; it is as if, someone besotted with the idea of getting across an ocean, difficult to cross, were to use an oar."

I would like to begin by mentioning briefly some of the pervasive values which come out of Anderson's life and work as a condensed matter physicist, but have and will continue to have crucial effects far beyond as well. The first of these, namely emergence, is encapsulated in the slogan "More is Different", a rallying cry which has empowered a large number of activities and persons. Two others, namely his observations on the (experimental) moorings of modern science, and on the essentiality of mechanism in scientific explanation, are believed in by a large segment of the community in this field. All three are essential not only for the good health and vigor of condensed matter physics, but of all science. As the noise dies down, I think these will, consciously or unconsciously, invigorate science in the future. Then, out of Anderson's more than five hundred published contributions, I mention a few seminal ideas briefly.

1. Emergence

While Physics is largely identified in the minds of both physicists and non-physicists with the "Great Game" (Kipling's phrase) of reductionism with its amazing and continuing triumphs especially in the realm of the very small, a remarkable feature of the last half century is the emergence of emergence, largely through our recognition of many instances of it. This is a growth industry in Physics, exemplified there in many explicit ways in the subfield of condensed matter physics. For example, Steven Weinberg, the famous particle physicist whose foundational creative work helped established our current picture of the very small, describes the BCS theory of superconductivity (1956–1957) thus: "I think the important thing accomplished by the theory of Bardeen, Cooper and Schrieffer (BCS) was to show that superconductivity is not part of the reductionist frontier." Anderson's long professional life has a few major themes running through it, e.g., broken symmetry, effects of disorder, of strong correlation, and of complexity, all of which illustrate emergence vividly.

Some sentences from the vastly influential short 1972 paper in Science by Anderson, called "More is Different" are the following. He says: "The ability to reduce everything to simple fundamental laws does not imply the ability to start from these laws and reconstruct the universe." "Instead, at each level of complexity entirely new properties appear, and the understanding of new behaviors requires research which I think is as fundamental in its nature as any other." "The constructionist hypothesis breaks down when confronted with the twin difficulties of scale and complexity."

It is not that emergence is a new or unknown principle. It is a term widely used for example in Biology, starting with the 19th century. In modern parlance, it is the "God principle". As Anderson says, 'the world in which we live is the consequence primarily ... or only — not of some incredibly simple but hidden "God equation" or "God particle" but of the "God principle."

The implication of the new emergentist view is felt well outside science. For example, a recent (2006) PhD thesis in Philosophy at Oxford (Paul Mainwood, Merton College) is entitled "More is Different? Emergent Properties in Physics." I quote the introductory paragraph:

> "This is an assessment of a contemporary movement, influential among physicists, about the status of microscopic and macroscopic properties. Although it is a recognizable version of older metaphysical theses of emergentism, these "New Emergentists" support their position by appealing to recent discoveries in condensed matter physics. The fountainhead for the movement is a short 1972 paper "More is Different," written by physicist Phillip Anderson. Each of my chapters is concerned with themes mentioned in that paper, or subsequently expounded by Anderson and his followers." While most physicists will not be positively affected by recognition from this quarter, the burden of the thesis is to argue for Emergence as a *metaphysical* position.

Ideas such as collective behavior and collective excitations, broken symmetry, "protection", new organizing principles at different levels of complexity, have all taken root as things which work, describing natural phenomena. We are all, all of us scientists, emergentists as well as reductionists though the latter flavor is more evident. The relative content of one or the other obviously varies from one scientist to the other, depending on the scientific discipline, on the subfield in that discipline, the time, and the influences. But science as pure reductionism is less and less tenable; there is increasing realization of the support for "emergentism." Many of us may be "emergentists" without knowing it, much like the character in Moliere's play who did not know that what he was speaking was prose. Constructionism starting from elementary laws seems to be a matter of faith, with a diminishing number of adherents. There seems to be a large amount of empirical evidence against it, though it may be a soothing thought for some of us.

It seems to me that if physics is to survive and flourish as a living, vital part of science, it has to not merely accept but to propagate the essentiality of emergence and to integrate it in theory and practice. Maybe this is happening even without our explicitly saying so; the way broken symmetry has informed the world of elementary particles and interactions is an old example. There is sustained work on the emergence of gravity via the realization that Einstein's equation for gravity can be cast in the image of thermodynamics. Further afield, there is speculation that consciousness might be the ultimate emergent phenomenon.

2. The Moorings of Modern Science

Anderson says in the beginning of a recently published review by him that "over four centuries ago, Francis Bacon, in his *Novum Organum*, outlined the philosophy which came to be the distinguishing characteristic of modern science. The philosophy held that knowledge of the nature of things was to be gained by the acute observation of nature, not by the study of authoritative texts or holy books. The resulting explosive growth of our understanding of the universe and of our ability to manipulate it cannot be gainsaid; whatever one may say about the technical ingenuity of the medieval Chinese or early mathematical discoveries of the Indians and the Arabs, one has to concede that nothing remotely resembling modern systematic science developed in those cultures". Of course there is the Cartesian view that knowledge is created by the human intellect, and not by observation (Einstein, the greatest scientist of the last century, said "it is the theory which describes what we can observe"). However, I think we all believe that observation and controlled experiment are the wellsprings of modern science.

As Anderson has emphasized, these as well as formulation of a hypothesis and its precise description, predicting new consequences, changing the theory as needed (thus acquiring firm and generally useful knowledge of the workings of the natural world in the process) are a continuous network. Sometimes, and specially in physics, the length and "tortuosity" of the links, and the distance between theory and experiment varies. This can often give rise to the dangerous illusion that these are autonomous activities. Perhaps the provocative statement of the famous Russian mathematician Vladimir Arnold, that "Mathematics is a part of physics. Physics is an experimental science, a natural science. Mathematics is that part of physics in which experiments are cheap," is a good jolt. For those of us (such as me) who are flummoxed by the last statement, Arnold added: "The Jacobi identity (which forces the heights of a triangle to cross at one point) is an experimental fact in the same way as that the Earth is round, but can be discovered with much less expense."

3. Mechanism

Anderson has always emphasized that knowing and describing the mechanism of a phenomenon is the key to its scientific explanation. It is natural to his conviction of the experimental underpinnings of science as well as his Bell Labs growth. This emphasis needs to be particularly strong in a field filled with strange phenomena. An eminent elementary particle physicist mentions for example that condensed matter physicists are often motivated to deal with phenomena because the phenomena themselves are so interesting. (He contrasts it with the situation in his field.) This variety in emergence obviously leads to a natural search for an appropriate mechanism, with general structures and universalities kept in mind. For example, in a recent review of Laughlin's book "*A Different Universe*: *Reinventing Physics from the bottom up*" (strongly influenced by "New Emergentism") he says, "Was it Pierre Weiss in 1905, with his mysterious molecular field and Weiss magnetons or Heisenberg, with the quantum theory, who explained ferromagnetism? I think the latter."

4. Some Pickings

It is impossible to indicate the full impact of Anderson's work by choosing a few ideas. The contributions not only consist of 540 papers (and still going strong!), but also of his influence through often game changing contact with several generations of condensed matter physicists. There is a very large number of original contributions; even a purely statistical analysis (published in *Physics World*, Aug. 17, 2006) declares him to be the most creative physicist in the world. I mention here a few pickings: these are obviously limited not by space, but by my own understanding and prejudices. The broad theme of emergence runs through them; broken symmetry, novel effects of disorder, strong correlations and complexity lead to new phenomena whose essential aspects are reflected in some of the choices below. I am sure that many physicists will miss several great new ideas brought to their awareness by Anderson. Two level systems, soft modes in ferrolectricity, gauge invariance as well as impurity effects in superconductors, ABM phase in ^{3}He, occurrence of multiple ionization states in semiconductors and frustrated magnetism are only a few of the possibilities.

(i) *Antiferromagnetism*

Long range ordered Néel antiferromagnets with equal spins in the two sublattices pointing in opposite directions began to be experimentally explored in the fifties via neutron scattering. Anderson's 1952 paper (*Phys. Rev. B* **86**, 694 (1952)) recognizes that antiferromagnets constitute a new kind of broken symmetry; one in which the ground state is not even an eigenstate of the Hamiltonian (unlike ferromagnetism where the ground state is one of the degenerate eigenstates of the Hamiltonian, and like superconductivity for which a good, revolutionary ground state wave function was written down a few years later by Bardeen, Cooper and Schrieffer or BCS). In both ferromagnets and antiferromagnets the ground state breaks the full symmetry of the Hamiltonian, and there are the characteristic, inevitable, Goldstone modes (spin waves). Anderson showed that because the Neél ground state is not an eigenstate, there is an inevitable dynamical symmetry restoration, described the relevant effective Hamiltonian (of the entire long range ordered AF) and estimated the time scale for this symmetry restoration to be $\sim$ years for a typical macroscopic Neél antiferromagnet. In most systems, there are small terms (e.g., single site anisotropy) which "pin" the Néel order. Anderson says of this paper: "It contains the seeds of *all* my further ideas about broken continuous symmetries and Goldstone modes, and the relation between microscopic and macroscopic physics."

(ii) *Anderson localization*

The second idea due to Anderson, perhaps the most important specific one in terms of its consequences in science, is in the paper (*Phys. Rev.* **109**, 1492 (1957)) which saw the proposal of Anderson localization, a phenomena generic to all waves in random media. It is the spatial localization of waves diffusing in disordered systems for sufficiently strong disorder. The motivational context, as repeatedly mentioned by Anderson, was the observation of Feher and co-workers of the apparent localization (non-diffusion) of P spins in lightly doped $Si{:}P$ where the n-type substitutional P impurities are randomly located in the Si matrix. He proposed and analyzed an even simpler model of spinless particles hopping quantum mechanically from site to site; the randomness is provided by the site energies. By a careful analysis of the probability distribution of the single particle propagator, he found that if the disorder exceeds a certain amount, there is a nonzero probability that the quantum particle stays home, i.e. is localized and does not propagate diffusively. Technically the new idea is based on the fact that in disordered systems, probability distributions (naturally random) are important, not just averages. At least two reasons why localization is unexpected are that the diffusion does not simply decrease continuously as disorder increases but *ceases* at a critical disorder, and that there is no "potential well" present that localizes the quantum particle. Though the phenomenon was identified in the context of quantum particles, (de Broglie waves with typically atomic wavelengths) it is common to *all* waves in disordered media; it has for example been argued that oceanic Rossby waves (with characteristic wavelength of order tens to hundreds of kilometers) can be Anderson localized, while the original discussion was for electron waves with atomic wavelengths. The 1979 Nobel prize awarded to Anderson cites this as one of the two contributions leading to it, the other being his work on localized magnetic moments in metals, mentioned below.

(iii) *Superexchange*

Insulators with magnetic ions are overwhelmingly antiferromagnetic. Ferromagnets are preponderantly metallic. Exchange due to identity of electrons, as first pointed out by Heisenberg, is ferromagnetic, and associable with electron motion or metallicity. So then, why is antiferromagnetism so common, and in magnetic insulators? The broad idea, due to Kramers (*Physica* **1**, 182 (1934)) is that in unfilled d-shell transition metal ion insulators with an intervening filled shell (diamagnetic) ion (e.g., Mn^{++} O^-Mn^{++} is a typical arrangement) there is a new phenomenon called superexchange. In this, while the d-shell ions are too far for their wave functions to overlap much directly, the presence of the bridging oxygen enables a "superexchange" process to occur, involving an intermediate state whose energy is high because of correlation effects. Anderson, in sustained work extending over nearly a decade, starting from 1950 and evolving to final form in 1959 (*Phys. Rev.* **115**, 2 (1959)), worked out the actual mechanism in detail. He showed that the resulting term was indeed of the exchange form $J_{ij}\vec{S}_i \cdot \vec{S}_j$, obtained J_{ij} for different situations and confronted theory with experiment regarding the actual magnetic structure (being unveiled by neutron scattering), as well as the Néel temperatures and their systematics. The realistic calculation of the kinetic and potential exchange terms for a number of systems includes not just extensive solid state chemical physics, but is perhaps the first recognition of the extent of the Mott phenomenon and its connection with the magnetic insulating state (a well-known lattice model due to Hubbard captures this correlation effect). The semiempirical Kanamori–Goodenough rules were also given a theoretical foundation. The Kramers–Anderson superexchange is now a staple of solid state materials science, e.g., the magnetic properties of ferrites, transition metal containing oxides, fluorides.... A part of the early post Second World War phase of solid state physics, this heroic creative effort (constrained and directed by experiment) is mainly a memory now, along with things like effective mass theory of semiconductors.

(iv) *Local magnetic moments in a metal*

In 1960, puzzling results were published by Matthias and co-workers, highlighting the fact that magnetic atoms (e.g., Fe, Co, Ni) when dissolved at low concentration in metals, were magnetic or not depending on a continuously changing property (e.g., electron density) of the solvent metal. Anderson proposed a simple, now canonical, model of a "magnetic" impurity atom in a metal. The former has the number of atomic electrons essentially constrained to an integral value (such that it has a nonzero magnetic moment) by strong Coulomb repulsion. The host metal consists of free conduction electrons. There is locally, a quantum mechanical hybridization between the electronic states of the magnetic impurity and of the metal. This simple Anderson model (*Phys. Rev.* **124**, 41 (1961)) exhibits, in a self-consistent mean field (Hartree–Fock) solution, a magnetic state if the hybridization strength is less than that of local correlation, otherwise not. This is the qualitative reason for the experimental observations. The Anderson model, with many concessions to reality (e.g., many orbitals, crystal fields, many "impurities" forming a lattice, namely the Anderson lattice) is alive and kicking. While related ideas connected with the nature of impurity states which hybridize with conduction electrons (resonant level) were in the air (especially through the work of Blandin and Friedel), the Anderson model captures the Mott metal insulator transition like dichotomous fate of a magnetic atom embedded in a simple metal. It is a standard part of modern of condensed matter physics. Its inevitable disappearance at exponentially low temperatures (the Kondo effect) is an interesting story, see below (items (vi) and (viii)).

(v) *Plasmon, gauge invariance and mass*

The Higgs mechanism, in which a broken symmetry massless gauge boson acquires mass, is due to Anderson (*Phys. Rev.* **130**, 439 (1963)). Nambu (whose work on gauge invariance and broken symmetry is pioneering) and Higgs are among those who have pointed this out. Anderson showed this quite explicitly in a non-relativistic field theoretic model in which the photon, a massless gauge boson, acquires mass by coupling to nonzero electron density. The massless phase mode or φ mode of the superconducting order parameter bosonic field $\Delta \exp(i\varphi)$, coupling to the massless photon mode, disappears and the photon acquires the plasmon mass while the amplitude fluctuations (fluctuations in Δ) of the bosonic field (Higgs field) are massive (the Higgs boson). This is the "signature" of the Higgs boson. This eventuality and several others whereby the massless modes in different broken symmetry condensed matter systems acquire mass are described in Anderson's paper which also indicates the relevance to a central problem of relativistic quantum field theory. In the condensed matter context where a dynamical theory of this broken symmetry is available, the experiments of Sooryakumar and Klein (*Phys. Rev. Lett.* **45**, 660 (1980)) and the explicit theoretical analysis of Littlewood and Varma (*Phys. Rev. Lett.* **47**, 811 (1981)) show this mass to be very nearly 2Δ where Δ is the magnitude of the superconducting gap. The Higgs boson, which via its coupling to fermion fields is the basic universal mass generator, was identified in the experiments at the Large Hadron Collider in CERN in 2012 to have a mass $\sim$125 GeV. While the history of the Higgs boson has been written about extensively, especially recently, and the future of the Higgs mechanism in particle physics has been speculated upon, the Higgs boson in the condensed

matter context (with known quantum dynamics) has been much less inves-
tigated. This is an example of the undated and expansive character of this
work.

(vi) *Orthogonality catastrophe*

Anderson's realization (*Phys. Rev. Lett.* **18**, 294 (1967)) that the
overlap between the ground states of a free electron gas with and
without a local electron scattering potential is exponentially small,
$\sim \exp[-(\ln N) \sum_l \frac{(2l+1)}{3\pi^2} \sin^2 \delta_l]$ where l is a positive integer and N is the
number of electrons in the system, and that this happens because of the
high density ($\propto$ energy) of low energy electron hole excitations in the Fermi
gas, seems, superficially, an esoteric if cute phenomenon. It is directly relevant to the observed power law lineshapes in spectra involving the fast
ejection and absorption of electrons locally; but that may also seem an interesting curiosity. But the fact that it is central to all low energy processes
involving local time dependent changes in a Fermi gas makes it an important basic feature of Fermi systems. It is at the root of the Kondo effect
in which the local spin changes its state because of coupling to the electron gas (the metal). This again was realized by Anderson, and is the basic
physical ingredient of the Kondo effect; solutions of this recalcitrant theoretical problem (e.g., Anderson and Yuval) are based explicitly or implicitly
on this.

(vii) *Spin glasses*

An obscure phenomenon, namely a sharp cusp in the temperature dependent ac magnetic susceptibility of relatively dilute magnetic alloys (e.g., Mn
in Au) was nailed down in the early sixties. The temperature at which this
occurs is found to be proportional to the concentration of the magnetic
atoms, clearly pointing to the (pairwise) interaction between them as the
cause. In a model where the interaction between them is mediated only by
the host conduction electron gas which has an oscillatory RKKY term coupling the two magnetic impurities the interaction can be ferromagnetic or
antiferromagnetic depending on the distance. The magnetic interaction is
thus "frustrated" due to the random location of the magnetic atoms. In general, one can model the random system as a collection of interacting spins
with a distribution of ferro and antiferromagnetic interactions. Initially, it
was believed that in such a collection, the spin orientations will freeze in
random directions because of the interaction when the system is cold enough

(hence the name spin glass), and early approaches did not think of this as a phase transition. In work of explosive significance, Edwards and Anderson (*J. Phys. F* **5**, 965 (1974)) proposed that the spins freeze in random directions but at a single temperature. They identified an order parameter which is the overlap between spin orientations at infinitely far separated times. This order parameter is nonzero below the spin glass transition temperature. The transition has clear equilibrium signatures, e.g., the measured second derivative of the magnetization with respect to the magnetic field diverges there. That it is a spin freezing transition is indicated e.g., by the appearance of a magnetic field caused by the frozen spins and felt by a muon injected into the system. Phenomena such as hysteresis and time-dependent remanence are other indicators.

Describing the transition and the frozen phase has led to new ideas such as replica symmetry breaking (Parisi), ultrametricity (Mezard and co-workers), as well as to sharpening of old ones such as free energy landscapes and non-ergodicity. Far more importantly, large and apparently unrelated problems in computer science, mathematics, neural networks and memory, protein folding, econophysics, indeed in all situations involving random, frustrated and competing configurations or possibilities have been illuminated by the journey started by Edwards and Anderson. Simulated annealing, directly inspired by spin glass theory, and unimaginable as a method, has proved a godsend for exploring otherwise inaccessible configurations in optimization problems; the occurrence of a phase transition and the "localization" of a random complex system in one of a large number of minima separated by (classically) impassable barriers is an idea which is an obvious consequence of the spin glass model, one whose effects have been widely felt in otherwise inexplicable ways. We see here clearly the strange consequences emerging from complexity.

(viii) *Poor man's scaling and the Kondo effect*

In the sixties interest was revived in an obscure (but troubling) phenomenon noted in the thirties, namely a maximum in the electrical resistivity of dilute magnetic alloys (*Physica* **1**, 1115 (1934)). Kondo noted that (*Prog. Theor. Phys.* **32**, 37 (1965)) the (weak, antiferromagnetic) coupling J between the magnetic moment of the impurity and the conduction electrons, perturbatively treated, produces a logarithmically diverging contribution to resistivity as temperature decreases. This unleashed the Kondo effect, a subject which occupied a good fraction of the quantum condensed matter community in the sixties and seventies. Two of Anderson's lasting contributions to physics (this and item (vi)) are connected with this.

Anderson noted, in perhaps (?) the first application of such scaling ideas in condensed matter physics, that the effect of eliminating the high energy electronic excitations is to strengthen J which is thus a scale dependent quantity tending (effectively?) to infinitely strong coupling in the limit of low energies and leading to a Fermi liquid fixed point. The ideas of renormalization, scaling and fixed points are extraordinarily fruitful; their application to continuous phase transitions and to a detailed (numerical (?)) analysis of the Kondo problem by Wilson have totally transformed these fields and have deep implications for systems with a large range of spatial and temporal degrees of freedom. In the Kondo case, this effectively very strong AF coupling leads to the disappearance of the moment. This is the ultimate fate of localized moments in metals, whose character was also first explored by Anderson (item (iv) above). The analogy with strong interactions which are also scale dependent, and in which the constitutents are relatively free quarks (i.e. like local moments) at high energies (temperatures) but strongly bound into at low energies (local moments bind into non-magnetic singlets) is deep.

In the realm of Kondo physics in condensed matter, the vanished moment leaves behind a characteristic Fermi liquid at low temperatures, which is accessed for example via a novel Fermi liquid effective Hamiltonian determined by scaling to strong coupling. The nature of the ground state and excitations for arbitrary spin and other real life conditions has been an obvious by-product.

(ix) *Weak localization*

It took a while to appreciate the nature and implications of Anderson's seminal theoretical discovery of localization, now more than half a century old. The first decade ($\sim$1957–1967) was a relatively fallow period. In the second, Sir Nevil Mott, more than others, worked out implications such as the mobility edge, Anderson metal insulator transition, and the electrical transport properties of Anderson insulators (or Fermi glasses) determined by a novel "variable range" hopping. He realized that electron localization due to disorder was the cornerstone for the physical properties of an entire large class of materials. He also argued that a metal with static disorder has to have a conductivity more than a minimum value. By the seventies, the field was lively again. Thouless proposed a new, reactive, diagnostic for localization based on the sensitivity of single particle energies of finite sized electronic systems to boundary perturbation. This introduced the idea of "Thouless conductance". In the late seventies, Anderson, with Abrahams, Licciardello and Ramakrishnan (*Phys. Rev. Lett.* **42**, 673 (1979)) proposed an approach to localization based on the behavior of conductivity as a function of scale size of the system, related to the scaling of the Thouless conductance. This calculation of conductivity as reduced by a specific quantum mechanical interference process (via "maximally crossed" diagrams) for the first time suggested a mechanism for the onset of localization as disorder increases, and led to predictions which could be compared in detail with measurements. The approach also states that there is no metallic ground state for non-interacting electrons moving in two dimensions in a random medium. This theory of *weak* localization, in its self-consistent form, is believed by some to be the (sole) mechanism of Anderson localization; it is not certain that this is indeed the process by which electronic states are localized in high dimensions too (strong localization?); for example a self-consistent, single particle propagator theory by Abou-Chacra, Thouless and Anderson (*Journal of Physics C: Solid State Physics* **6**, 1734 (1973)) for the Bethe lattice in the early seventies looks very different. This work has been very consequential. Investigations of localization of light, of sound and of matter waves are well known spinoffs. The idea of electrical current or flow of electrons in a wire being a quantum process unlike say the flow of water in a river, has taken root, and has been a major stimulus in our understanding of electrical transport in meso- and nanoscopic systems where quantum effects and the quantum conductance scale ($\frac{2e^2}{n} \simeq (12.9\,\mathrm{k\Omega})^{-1}$) are prominent.

(x) *RVB and high-temperature superconductivity*

I end by briefly noting the major interest and contribution of Anderson in the last quarter of a century, namely his proposal of a specific strong correlation mechanism (the resonating valence bond or RVB mechanism) for the occurrence of high-temperature superconductivity in the cuprates. Very soon after its discovery in the late eighties (and after one of the very first public discussions of it in a "Mixed Valence" international conference in Bangalore, India) Anderson proposed (Science, **235**, 1196 (1987)) a theory for it. According to it, the relevant Copper *d* electrons form pairs (valence bonds); the superconductor is, electronically, a phase coherent collection of such spin singlet bonds of different bond lengths (RVB).

The RVB idea itself, originating qualitatively with Linus Pauling's description of the metallic state as one such, and given life by Anderson in the seventies, harks back to a concern as old as that of Landau (and of others) in

the thirties. The worry has been that for a pair of antiferromagnetically coupled spins, the ground state is a spin singlet; this is an unlikely parent for the
observed Néel ground state in antiferromagnets which presumably consist of
a lattice of spins with pairwise, short range antiferromagnetic superexchange
coupling. Indeed, Anderson himself had shown in the seventies that the RVB
"spin liquid" ground states are quite competitive energetically with (spin
solid?) Néel antiferromagnets. If indeed the spins pair into an RVB liquid,
why exactly is it not seen in the undoped cuprates, and how does hole doping lead such a state to become superconducting? Answers to these and
other questions have been proposed, and Anderson has consistently used
this picture (with additional hypotheses as necessary) to confront a number of facts successfully, including electron spectroscopic data. However, this
specific explanation is not universally accepted; the reasons are sociological
as much as scientific. We have become ensnared by the BCS theory with
phonon exchange as the pairing mechanism, and there is also a strong clash
of egos underlying the apparent scientific discourse.

The approach of Anderson clearly marks out a new path for electron
pairing to occur, namely one which involves (only) strong repulsive electronic correlations. This has led to many major new ideas; the field of strong
electron correlations in metals is vigorously active.

The RVB idea has also focused attention on quantum spin liquids. These
are not theoretical speculations alone. In many geometrically frustrated spin
lattices, there are clear experimental signatures for the occurrence of such a
state of matter. This last (and lasting) contribution of Anderson condensed
matter physics, while addressing the high-T_c phenomenon in a characteris-
tically imaginative and unfettered way, and respecting experimental reality,
has spawned the exploration of a large range of strong correlation, quantum
spin phenomena. It is an indicator of his creativity and courage.

I conclude with a statement from Anderson which appears in the introduction to the book containing selected papers by him (*A Career in
Theoretical Physics*, World Scientific, Singapore, 2nd Edn., 2004). The contribution of Physics is the method of dealing correctly both with the sub-
strate from which emergence takes place, and with the emergent phenomenon
itself.... Ever newer insights into the nature of the world around us will
continuously arise from this style of doing science.

I mark our admiration for the greatest exemplar of this style.

Superfluidity and Symmetry Breaking — An Anderson Living Legacy

Frank Wilczek

Center for Theoretical Physics, MIT
Cambridge, MA 02139, USA
wilczek@mit.edu

This is an eclectic survey of concepts around superfluidity and symmetry breaking, prepared for the celebration of Phil Anderson's 90th birthday in October 2013. I emphasize, through major examples, that the concepts Anderson pioneered in this field have very wide scope, penetrating in particular into many central issues of high energy physics: electroweak symmetry breaking, confinement, chiral symmetry breaking, and Majorana mass. I also illustrate how Anderson's pseudospin method can be used to exemplify breaking of time translation symmetry.

Phil Anderson doesn't present himself as a high energy physicist, but he's been a very influential one. In particular, his focus on *symmetry* and its breaking, and the clarity of vision he achieved through his analysis of concrete problems in magnetism, superfluidity, and superconductivity, have helped us to see many other things more clearly, and in new ways. We've come to appreciate that we live inside a cosmic superconductor.

The theory of symmetry breaking has two main aspects: macroscopic, and microscopic. The macroscopic part centers around the consequences of symmetry breaking, the microscopic part centers around the mechanisms that cause it. Anderson made major contributions to both parts.

Here, by way of tribute, I will discuss three things.

- First, I will present the distilled essence of global and local symmetry breaking in two simple, canonical models. Those models arose by abstraction from superfluid ^{4}He and standard superconductors, respectively. They capture central aspects of the macroscopic theory, and their ideas carry over to chiral symmetry breaking in QCD and to electroweak symmetry breaking, respectively.

- Second, I will discuss the theory of hadronic matter at ultra-high density (large baryon number, low temperature). The microscopic pairing theory pioneered by BCS, which Anderson both clarified and generalized, is beautifully adapted to this problem. It suggests a form of symmetry breaking that leads, through the macroscopic theory, to some very striking results. In particular, the classic "hard" non-perturbative challenges of QCD, to demonstrate confinement and chiral symmetry breaking, are consequences.
- Finally, I will briefly mention two new applications, that I've been thinking about recently. I'll save the details for the main text; here I'll just say that they demonstrate the continued vitality of Anderson's ideas and style of thinking, as we bring them to bear on new questions.

Anderson's work on the renormalization group has also had profound impact in high energy physics, but I will not discuss that here.

1. Canonical Models: Superfluidity and Superconductivity

1.1. *Superfluidity: Symmetry breaking*

The simplest model for superfluidity involves a complex scalar field that supports a phase ($U(1)$) symmetry in its fundamental equations, but not in their stable solutions. This sort of theory describes the superfluidity of liquid ^{4}He. In that context, the scalar field creates and destroys helium atoms, and the (spontaneously broken) phase symmetry is associated, through Emmy Noether's famous theorem, with conservation of ^{4}He atom number.

The scalar field is subject to a potential

$$V(\phi) = -\frac{\mu^2}{2}\phi^*\phi + \frac{\lambda}{4}(\phi^*\phi)^2. \tag{1}$$

The energy, as a function of the expectation value of the field, is minimized at

$$|\langle|\phi|\rangle| = v \neq 0 \tag{2}$$

$$v = \frac{\mu}{\sqrt{\lambda}} \tag{3}$$

and to minimize the kinetic (gradient) energy we must choose a single space-time constant value for $\langle|\phi|\rangle$. Whereas the potential, and the equations of motion of the theory as a whole, are left invariant under multiplication of the ϕ field by an arbitrary phase factor, those specific minimum energy solutions are not.

Deep issues arise around the precise realization of candidate states that support solutions of Eq. (2), in view of the fact that ϕ is subject to quantum and thermal fluctuations. The strength of the coupling, the dimensionality of the system, and its volume are all relevant factors. Phil Anderson contributed greatly to our understanding of those issues, but that belongs to a branch of his work (the renormalization group) that I won't pursue here. The basic upshot is that when the coupling is sufficiently weak, the temperature sufficiently low, the dimension sufficiently high, and the system sufficiently large — and only then! — we can justify the naive procedure.

To understand the consequences of Eq. (2), we expand around a typical one, where the expectation value is simply $\langle \phi \rangle$, with no phase factor, according to

$$\phi \equiv (v + \rho)e^{i\theta} \equiv (v + \rho)e^{i\sigma/v}. \tag{4}$$

Here ρ and σ are fields, that vanish in the minimum energy state (ground state) we have chosen. We describe procedure as expanding around a condensate. Expressing the basic Lagrangian in terms of these fields, we have

$$\mathcal{L} = \frac{1}{2}\partial^\mu \phi^* \partial_\mu \phi - V(\phi) \tag{5}$$

$$\rightarrow \frac{1}{2}\left(1 + \frac{\rho}{v}\right)^2 \partial^\mu \sigma \partial_\mu \sigma + \frac{1}{2}\partial^\mu \rho \partial_\mu \rho - \tilde{V}(\rho) \tag{6}$$

$$\tilde{V}(\rho) = -\frac{\mu^4}{4\lambda} + \mu^2 \rho^2 + \frac{3\sqrt{\lambda}}{4\mu}\rho^3 + \frac{\lambda}{4}\rho^4. \tag{7}$$

The interpretation of Eqs. (6) and (7) is simple. The σ field occurs only in the first term of Eq. (6). Therefore it represents a soft mode: Its variations carry zero energy, in the limit of infinitely long wavelength and low frequency. In a particle interpretation, it represents the field of a massless particle. Modes of this kind often arise when one expands around solutions which have less symmetry than the underlying equations, and they are commonly referred to as Nambu–Goldstone bosons.[a] One also has non-linear interactions, between the σ field and the amplitude fluctuation field ρ.

The characteristic "superfluid" phenomenology of superfluid ^{4}He is due to the existence of this mode. The phase symmetry of the underlying equations, which is spontaneously broken in our ground state, was meant to encode conservation of ^{4}He atoms. Of course, conservation of the number of ^{4}He atoms is not really violated. Enclosing any finite sample of ^{4}He we can

---

[a]The naming of this, and of several other concepts I'll be discussing, itself involves a sort of spontaneous symmetry breaking. While never wholly inappropriate, the chosen names typically reflect historical choices among a number of valid possibilities, which have been frozen into the literature in order to minimize some sort of ambiguity function. Rectification of names is a profound task, and I will not attempt it here.

draw an enclosing surface, at which we can apply the reasoning that leads to conservation. But if we draw a surface within the sample we must worry about surface terms, and that changes the picture completely. The observable consequence of the mathematical violation, is that this quantum number is easily moved. That is a heuristic way to understand superfluidity: Superfluidity is flow mediated by the soft modes or Nambu–Goldstone bosons.

The ρ field, which parametrizes fluctuations in the amplitude of the condensate, in more conventional. Considering Eqs. (6) and (7) we see that it represents a hard mode, whose quanta are (quasi)particles of mass $\sqrt{2}\mu$. The cubic and quartic terms in Eq. (7) represent self-interactions of these quanta, and the numerical term $-\frac{\mu^4}{4\lambda}$ represents the energy density gained through condensation. When the theory is coupled to gravity, it represents a contribution to the cosmological term, that can be profoundly problematic (see below).

1.2. *Superconductivity: Gauged symmetry breaking*

Classic superconductors are described by a gauged version of the same model. This is a case of "More is Different," if ever there was one.

We introduce gauge field degrees of freedom using the Maxwell Lagrangian, and we replace ordinary by gauge covariant derivatives in the scalar kinetic terms. The potential is unchanged. Thus we consider

$$\mathcal{L}_{\text{kin.}} \rightarrow -\frac{1}{4}F^{\mu\nu}F_{\mu\nu} + \frac{1}{2}(\nabla^\mu \phi)^* \nabla_\mu \phi \tag{8}$$

$$F_{\mu\nu} = \partial_\mu A_\nu - \partial_\nu A_\mu \tag{9}$$

$$\nabla_\mu \phi = \partial_\mu \phi - igA_\mu \phi. \tag{10}$$

Expanding the covariant derivative term in Eq. (8) we find

$$(\nabla^\mu \phi)^* \nabla_\mu \phi \rightarrow \frac{1}{2}\left(1 + \frac{\rho}{v}\right)^2 (\partial^\mu \sigma - gvA^\mu)(\partial_\mu \sigma - gvA_\mu) \tag{11}$$

$$= \frac{g^2 v^2}{2}\left(1 + \frac{\rho}{v}\right)^2 \tilde{A}^\mu \tilde{A}_\mu \tag{12}$$

$$\tilde{A}_\mu \equiv A_\mu - \frac{1}{gv}\partial_\mu \sigma. \tag{13}$$

Once again the interpretation is simple, though at first encounter startling. It is natural, as in Eq. (13), to introduce a new field $\tilde{A}_\mu$, wherein the gauge potential A_μ is supplemented by a longitudinal contribution $-\frac{1}{gv}\partial_\mu \sigma$. The new field is still governed by the Maxwell Lagrangian, since the longitudinal field cancels in $F_{\mu\nu}$. But, according to Eq. (12), it has acquired mass

gv. We also find non-linear interactions between $\tilde{A}_\mu$ and the amplitude fluctuations ρ. The σ field, which represented a soft mode in the theory without gauge symmetry, no longer appears as a separate degree of freedom.

The characteristic macroscopic phenomenology of superconductors is largely implicit in this model, but subtle to extract. The major bulk feature of superconductors involves *removing* low-energy dynamics from the gauge field, as epitomized by the Meissner effect. One also induces an energy *gap* for the electrons, as I'll discuss momentarily. So the bulk signatures are essentially negative.

Reflecting this situation, on the formal side, we find that there is no proper bulk order parameter. The phase of the expectation value $\langle \phi \rangle$ is gauge dependent, and gauge dependent quantities are unphysical. Strictly speaking, the Hilbert space of the gauge theory must be restricted, or projected, onto gauge-invariant states; for only in that sector do we find a positive definite metric.

It is common to speak of "gauge symmetry breaking," but gauge symmetry actually represents a convenient mathematical redundancy in the physical description. Gauge symmetry must be factored out, and cannot be broken. We can and should speak instead of "gauged symmetry breaking". When the gauge coupling is weak, we can use the procedure above, based on adding the effect of gauge interactions to a model of global symmetry breaking, to draw approximate physical conclusions. That is the intellectually respectable interpretation of gauged symmetry breaking.

If we didn't get to view them "from the outside" — if we didn't have access to external electromagnetic fields, surfaces, and weak links — then superconductors would be featureless. Fortunately, we do! A vast wonderland, largely charted by Anderson, featuring the Meissner effect (as a *positive* phenomenon), persistent currents, Josephson effects, and many other phenomena both interesting and useful, opens before us.

1.3. *Application to electroweak symmetry breaking*

But if some form of intelligent life evolved within a superconductor, and experienced it only from the inside, the accessible manifestations of gauged symmetry breaking would be subtler. They might notice that there is a massive vector excitation (the photon) with a simple pattern of couplings, and wonder whether it might be described using a model based on gauged symmetry breaking. If the modeling went smoothly, and suggested new consequences that were verified, they might eventually convince ourselves that they live inside a superconductor.

Something very much like that has occurred over the last few decades of *human* history, culminating in the recent discovery of the Higgs boson.

The phenomenology of the weak interaction offered several clues that a description using gauged symmetry breaking would be appropriate. (Of course, this is much clearer in retrospect!) The $V-A$ current × current structure of the effective Hamiltonian, derived phenomenologically, begs to be explained by vector boson exchange. And the universal coupling strength for very different particles, suggests an underlying gauge invariance. A major difficulty, that held up the fruition of this idea by several years, was the apparent difficulty of reconciling gauge invariance with nonzero mass for the vector bosons. Anderson was the first to articulate that this difficulty might be only apparent, and he discussed several examples where it is transcended. Brout and Englert, Higgs, and also Guralnik, Hagen, and Kibble showed that it could be transcended at weak coupling and using Lagrangians well adapted to the needs of particles physics, along the lines we reviewed above. The underlying equations would look very familiar to Landau and Ginzburg, or even London, from their modeling of macroscopic superconductivity decades before; but it required an audacious psychological leap, to picture the *world* as a superconductor, *seen from the inside.*

Modern electroweak theory incorporates an extension of the classic model described above, and the Higgs boson is essentially the ρ field of that model. As such, it is the not the central player in the dynamics of spontaneous symmetry breaking, but an avatar. The same results for vector boson masses, specifically, would result if we had a different potential V, or if we had several scalar fields, keeping the appropriate weighted sum of contributions to the vector boson masses fixed. Nevertheless, in the spirit of Occam's razor — or its generalization, the Jesuit Credo that "It is more blessed to ask forgiveness than permission." — it makes good sense to analyze the consequences of the simplest possible model, to learn how well that approximates observed reality, if it does at all.

A picture of the discovery channel may afford the best introduction to what's involved (see Fig. 1).

Since the primary couplings of the Higgs particle to other fundamental particles in the standard model are proportional to those particles' masses, its direct, primary coupling of the matter our accelerator accelerates — protons — is highly suppressed. For protons mainly contain up and down quarks and antiquarks, whose mass is very nearly zero, and color gluons, whose mass is exactly zero. The main line of communication, as I discovered, is through quantum fluctuations. In the picture, we see that fact reflected in the virtual particle loop. Both gluons and quarks couple efficiently to

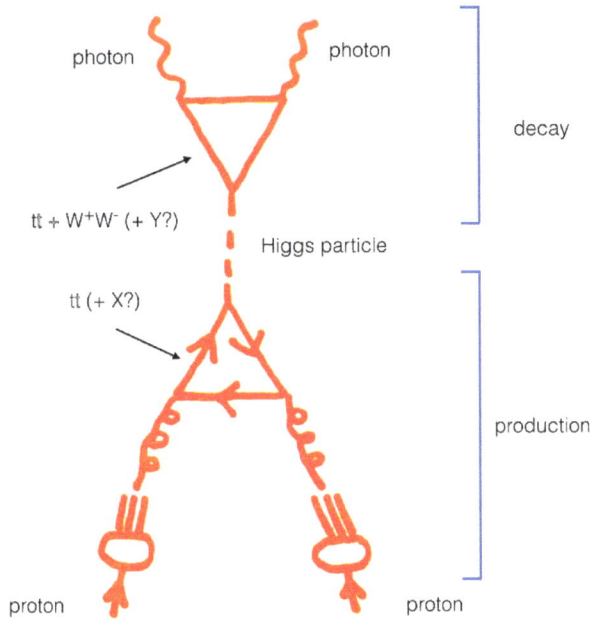

Fig. 1. This stylized Feynman diagram depicts the process through which the Higgs particle was discovered. As described in the text, it brings many aspects of our deep theory of matter into play. To follow the process in time, we advance from bottom to top. The beams at the Large Hadron Collider contain rapidly counter-moving protons, that are brought into collision. Protons are constructed, according to QCD, from more basic building blocks, including gluons, whose properties are simpler to analyze, and which (unlike holistic protons) support large energy-momentum flows. The core process, for Higgs production, involves collision of two gluons. (The remainders of the protons interact with one another and materialize in complex ways. They constitute a dominant "background," that one must control very well in order to a extract meaningful signal.) These connect to the Higgs particle, which then decays into two photons. Photon pairs with invariant mass in the range of interest are relatively difficult to produce by conventional standard model processes, so this signal rises above background.

virtual top quark — antitop antiquark ($t\bar{t}$) pairs: the gluons through their color charge, the Higgs particle through their mass. Once the Higgs particle is produced, it can decay into two photons through a similar process, this time involving either $t\bar{t}$ or W^+W^- virtual pairs. Although this is a small decay branch, it is very advantageous from an experimental point of view, because energetic photons are both quite distinctive and relatively difficult to produce in other ways.

This is what the discovery plot looked like Fig. 2 reproduces the discovery plot.

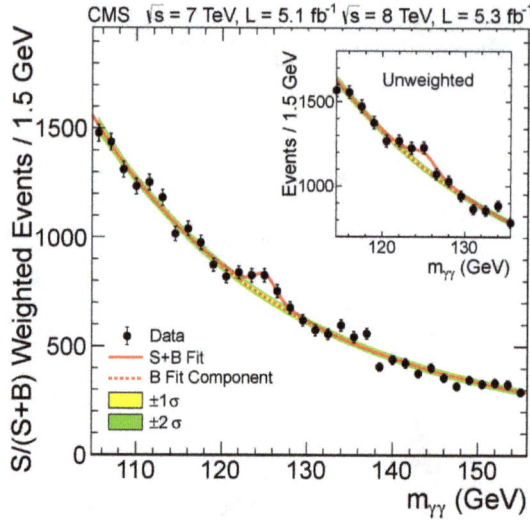

Fig. 2. This plot (courtesy of the CMS collaboration) of the production rate for two photons, as a function of their invariant mass, was among the first clear indications for the existence of a near-minimal Higgs particle, with mass close to 125 GeV. As you can see, it's very important to understand the background quantitatively! Thus the discovery confirmed the accuracy of the standard model, as a description of reality, and our ability to calculate its consequences, in many more ways than one.

Since the discovery announcement, the measurements have become both more extensive and more accurate. In particular, several additional production and decay channels have been explored.

So far, all results are consistent with the minimal model. It remains important to keep pushing, because by its nature the Higgs particle opens a portal into several not implausible forms of new physics. For example, considering the two photon channel, there could be contributions from heavier particles, yet unknown, to the virtual particle loops. Also, if the Higgs sector is not minimal — if there's more than one scalar field involved — the couplings of the observed particle will diverge from minimal expectations. It would be remarkable if the minimal model continues to survive close scrutiny, but I don't expect it.

1.4. *Conventional versus Majorana mass*

Before leaving this subject, I'd like to comment briefly on the use of gauged symmetry breaking to generate fermion masses. Here, there is a significant difference between the standard model and superconductivity. I will gloss over several technicalities, mainly highlighting one simple but important, entertaining point.

In the standard model, we move from couplings of the ϕ field to mass terms for fermions ψ basically as follows:

$$y\bar{\psi}\phi\psi \xrightarrow{\sim} yv\left(1 + \frac{\rho}{v}\right)\bar{\psi}\psi \equiv m_\psi\left(1 + \frac{\rho}{v}\right)\bar{\psi}\psi. \tag{14}$$

Together with the mass term, we have a proportional coupling to the ρ field.

In superconductors, the closest analogue to a mass term is the electron gap. It arises from electron-electron interactions, when we have pairing into a condensate, basically as follows:

$$\kappa\bar{e}\bar{e}ee \xrightarrow{\sim} \kappa\langle\bar{e}\bar{e}\rangle ee + \text{h.c.} \tag{15}$$

The condensate $\bar{e}\bar{e}$, in the microscopic theory, supplies the ϕ field of the macroscopic theory. But the mass term is of a different kind, involving particle number violation. Mass terms of this kind, though they do not arise in the standard model, are important in some of its possible extensions. They arise in the description of neutrinos, and in the description of various hypothetical particles that are introduced to implement supersymmetry. Particles whose mass is of this type can annihilate, at rest, in pairs. They are called Majorana fermions. We see that inside a superconductor, near the nominal Fermi surface, electron/hole quasiparticles are Majorana fermions.

(In metals the ordinary electron mass does not provide a gap, of course. Formally, that is because it is cancelled by the chemical potential.)

2. QCD Meets BCS

2.1. *Conceptual background*

A wise principle states "It is more blessed to ask forgiveness than permission." In that spirit, we consider the possibility of constructing a description of high-density QCD based on its elementary degrees of freedom, quarks and gluons.

Can we really bypass the forbidding complexities of nuclear physics, and use those elementary degrees of freedom directly? At first sight that idea looks extremely plausible. High density means large Fermi surfaces. Neglecting interactions, the low-energy excitations are associated with action near the Fermi surface: a mode just above the Fermi surface, empty in the ground state, becomes occupied, or a mode just below becomes empty. Since the Fermi surface is large, all the modes involved carry large momentum and energy. So scattering among these low-energy excitations will either involve only small angles, and leave the distribution nearly unchanged, or else bring in large momentum transfers, and therefore weak coupling (asymptotic freedom). It appears, therefore, that perturbation theory should work well; and perturbation theory is something we know how to do.

But when you actually do the calculations, you run into infrared divergences. They arise from two sources:

- The preceding argument only concerns the quarks. Its central point is that Pauli blocking removes the infrared divergences that usually arise through low-virtuality quarks. Gluons are not subject to any such effect. Color electric forces are screened by the quark medium, but magnetic forces remain long-ranged, and lead to infrared divergences.
- Interacting fermions are subject to the Cooper instability. One has many near-zero energy excitations at zero momentum, associated with particle-particle or hole–hole pairs with equal and opposite three-momenta $\pm\vec{p}$. Thus in perturbing around the free Fermi sphere one is engaging in highly degenerate perturbation theory. As a general matter, degenerate perturbation theory can result in significant restructuring of the ground state. In this specific context, Bardeen, Cooper, and Schrieffer (BCS) taught us that even a small attractive interaction *will* lead to a drastic re-arrangement of the ground state, by leading to pairing and superfluidity.

In conventional superconductors it is quite subtle to find an effective attractive interaction between electrons. The primary interaction between electrons is the Coulomb interaction, and it is of course repulsive. To find an attractive interaction one must bring in phonons, retardation, and screening, and concentrate on modes within a thin shell around the Fermi surface. For many "unconventional" superconductors, famously including the cuprates, the mechanism of attraction remains unclear. In all cases the superconducting transition temperature (which reflects the attractive dynamics) is far below the melting temperature (which reflects the primary dynamics).

In QCD it is more straightforward. The primary interaction can already be attractive. Two separated quarks, each in the triplet **3** representation, can be brought together in the antisymmetric **3̄**. The disturbance in the gluon field due to color charge is then half what it was; since the field energy has decreased, the force is attractive. By zeroing the spin — that is, once again, choosing the antisymmetric channel — we also remove the sources of magnetic disturbance. Thus on very general grounds we expect a powerful attractive interaction between quarks in the channel where both colors and spins are antisymmetric. This intuition is borne out by calculations using one-gluon exchange, instanton models, and direct lattice simulations.

Thus color superconductivity occurs straightforwardly and should be robust. What does it mean?

- Gluons acquire mass — that is a way to state the equations of the Meissner effect. If *all* the gluons acquire mass, their exchange will no longer produce infrared divergences.
- Quarks acquire mass — that is a way to state the equations of the energy gap. If *all* the quarks acquire mass, Cooper's infrared divergence will be removed.
- Therefore the weak-coupling expansion should work, as long as we start from the proper — color superconducting — ground state.
- This ground state does not contain massless gluons nor exhibit long-range forces. In that sense, it exhibits confinement. We also have the classic phenomenon of confinement — absence of fractional *electric* charge in the spectrum, as I'll explain shortly.
- The energy gap for quarks suggests that chiral symmetry, which is associated with massless quarks, may be broken.

In short, we have the prospect of a phase that exhibits the main non-perturbative features of QCD — confinement and chiral symmetry breaking — in a transparent, fully controlled theoretical framework. Let me emphasize that here I am speaking of a phase of QCD itself, not of some idealization of a model of a caricature of QCD.

Now let's see how all this is embodied, concretely, in equations.

2.2. *Ground state*

To bring out the central issues and ideas, I will assume as the initial default that all quarks are massless, that they are subject to a common chemical potential, and that electromagnetism can be treated as a perturbation. I'll circle back to revisit these assumptions, in due course.

Because the most attractive channel for quarks is antisymmetric both in color and spin, Fermi statistics requires another source of antisymmetry. One possibility is antisymmetry in the spatial wave function of the quark pairs. For example, we might have p-wave pairing. For simple, purely attractive interaction potentials, s-wave tends to be favored, because it allows pairs from all directions over the Fermi surface to act in phase. So s-wave pairing, if possible, is likely to be favorable.

The remaining possible source of antisymmetry is flavor. Thus we must pair off *different* flavors of quarks to take best advantage of the attractive interaction between quarks. This can bring in some significant complications. Obviously, it means that the one-flavor case is not representative, and that we cannot build up the analysis one flavor at a time. The two-flavor case also does not go smoothly. Antisymmetry in flavor and spin (and lack of orbital structure) reduces the quark-quark channel to a single vector in color space. Therefore condensation in this channel breaks color symmetry only partially, in the pattern $SU(3) \rightarrow SU(2)$. Some gluons remain massless, and some quarks remain gapless, so infrared divergences remain.

Simplicity and self-consistency (that is, consistent use of weak coupling) first arrive when we consider three flavors.

2.2.1. *Color flavor locking*

I'll describe the full structure of the condensate momentarily, but since that's a little intimidating let me begin with a sketch. Since the spin (singlet) and spatial (*s*-wave) structures are unremarkable I'll suppress them, and also chirality. The favored condensate should be antisymmetric in color and in flavor, which suggests the form

$$\langle q_a^\alpha q_b^\beta \rangle \sim \epsilon^{\alpha\beta*} \epsilon_{ab*}$$

where the Greek indices are for color, the Latin indices are for flavor, and $*$ is a wildcard. Now by setting the wild cards equal, and contracting, we maintain as much residual symmetry as possible. Any fixed choices for the wildcards will break both color and flavor symmetries. But by *locking* color to flavor we maintain symmetry under the combined (so-called diagonal) symmetry group. Thus we arrive at

$$\langle q_a^\alpha q_b^\beta \rangle \sim \epsilon^{\alpha\beta*} \epsilon_{ab*} \rightarrow \epsilon^{\alpha\beta i} \epsilon_{abi} \propto (\delta_a^\alpha \delta_b^\beta - \delta_b^\alpha \delta_a^\beta). \tag{16}$$

This condensate breaks local color times global flavor $SU(3) \times SU(3)$ to a diagonal, "modified flavor" global $SU(3)$. It also spontaneously breaks baryon number symmetry. To a particle physicist encountering these ideas for the first time, that might sound dramatic — and it is — but not in the sense of allowing the material to decay. With the sample enclosed in a finite volume, outside of which the order parameter vanishes, there is a strict conservation law for the integrated baryon number. As in the theory of liquid helium-4, where one speaks of a condensate of helium atoms, the true implication is that there is easy transport of baryon number within the sample. More specifically, there is a massless Nambu–Goldstone field, which supports the supercurrents characteristic of superfluidity.

Now comes the full structure, in all its glory:

$$\langle 1|(q_a^\alpha)_L^i(\vec{k})(q_b^\beta)_L^j(-\vec{k})|1\rangle$$

$$= \epsilon^{ij}\left(v_1(|\vec{k}|)(\delta_a^\alpha\delta_b^\beta - \delta_b^\alpha\delta_a^\beta) + v_2(|\vec{k}|)(\delta_a^\alpha\delta_b^\beta + \delta_b^\alpha\delta_a^\beta)\right)$$

$$= -(L \leftrightarrow R).$$

Some further words of explanation are in order. The mid-Latin indices i, j are for spin. The "L" and "R" are for left and right chirality. The relative sign between left and right condensates reflects conservation of parity. The functions $v_1(|\vec{k}|), v_2(|\vec{k}|)$ are, for weak coupling, peaked near the Fermi surface. Our preceding discussion anticipated v_1, but v_2 is also allowed by the residual symmetry. It emerges from calculations based on the microscopic theory, though with $v_1 \gg v_2$.

2.2.2. Symmetry breaking and symmetry transmutation

Tracking chiral flavor symmetry and baryon number together with color, the implied breaking pattern is:

$$SU(3)_{\text{color}} \times SU(3)_L \times SU(3)_R \times U(1)_B \to SU(3)_\Delta \times Z_2. \qquad (17)$$

The residual $SU(3)_\Delta$ global symmetry, and the Z_2 of fermion (quark) number, can be used to classify excitations. There is no residual local symmetry: all the color gluons have acquired mass. A slightly more refined analysis shows that all the quarks have acquired gaps.

Finally, as a consequence of the underlying (spontaneously broken) baryon number and chiral symmetries we also have the generalized ground states

$$\langle \mathbf{U},\theta|(q_a^\alpha)_L^i(\vec{k})(q_b^\beta)_L^j(-\vec{k})|\mathbf{U},\theta\rangle$$

$$= \epsilon^{ij}e^{i\theta}\left(v_1(|\vec{k}|)(\mathbf{U}_a^\alpha\mathbf{U}_b^\beta - \mathbf{U}_b^\alpha\mathbf{U}_a^\beta) + v_2(|\vec{k}|)(\mathbf{U}_a^\alpha\mathbf{U}_b^\beta + \mathbf{U}_b^\alpha\mathbf{U}_a^\beta)\right)$$

$$= -(L \leftrightarrow R)$$

for an any $SU(3)$ matrix $\mathbf{U}$. Low-frequency, long-wavelength modulation of the fields θ and $\mathbf{U}$, which represents slow motion within the vacuum manifold, generates the Nambu–Goldstone bosons.

Before leaving the ground state, one last comment. Throughout this discussion I've used the language of gauge symmetry breaking and gauge non-singlet order parameters. This is quite familiar and traditional in BCS theory, and also in the standard model of electroweak interactions. Strictly speaking, however, it based on lies, for local gauge invariance is never broken, and gauge-variant expectation values always vanish. Indeed, the physical Hilbert space is defined by restricting to gauge-invariant states. The usual procedures are a tool — a way of implementing favorable correlations in weak coupling. Their physical content emerges when we use them to draw consequences for gauge-invariant quantities, such as the physical spectrum or expectation values of gauge-invariant operators. In the CFL, we can identify two nonzero gauge invariant vacuum expectation values that break chiral or baryon-number symmetries, of the types:

$$\langle q_L q_L \bar{q}_R \bar{q}_R \rangle$$
$$\langle qqqqqq \rangle$$

with the color indices suitably contracted. They arise as powers of the primary condensates. (Including instanton effects, we also get $\langle q_L \bar{q}_R \rangle$.) By contrast conventional s-wave spin-singlet BCS condensation, and also doublet condensation in the standard electroweak model, do not support true order parameters.

2.2.3. Elementary excitations

We can analyze the elementary excitations from the point of their spin and quantum numbers under the residual $SU(3)_\Delta$ symmetry. There are three types:

1. *Excitations produced by the quark fields*: They are spin-$\frac{1}{2}$ fermions that decompose as $\mathbf{3} \times \bar{\mathbf{3}} \to \mathbf{8} + \mathbf{1}$ under $SU(3)_{\text{color}} \times SU(3)_{\text{flavor}} \to SU(3)_\Delta$. The singlet turns out, at weak coupling, to be significantly heavier than the octet.
2. *Excitations produced by the gluon fields*: They are spin-1 bosons that form an octet.
3. *Collective excitations*: They are a pseudoscalar octet of Nambu–Goldstone bosons, plus the singlet superfluid mode.

Overall, there is a striking resemblance between this calculated spectrum of low-lying excitations and what one might expect for the elementary excitations in the "nuclear physics" of QCD (that is, the nuclear physics of QCD with three massless flavors), based on standard phenomenology and modeling. The calculated elementary excitations map nicely onto the entries in the expected hadron spectrum. Even the superfluid mode makes sense, because we would expect, in this "nuclear physics", pairing in the dibaryon channel.

2.2.4. *Quark hadron continuity*

Since conventional (heuristic) "nuclear physics" and the asymptotic (calculated) CFL state match so well with regard both to their ground state symmetry and their low-lying spectrum, it is hard to avoid the conjecture that there is no phase transition separating these states. Consider cranking up the chemical potential, starting from zero. First there's Void. At a critical value, nuclear matter appears, with a first-order transition. After that, there's just smooth evolution.

This conjecture of *quark-hadron continuity* is both (superficially) paradoxical and powerful in implication.

It may seem paradoxical, when we consider that the baryons of conventional "nuclear physics" are supposed to go over smoothly into excitations produced by directly by single quark fields. After all, baryons are famous for containing three quarks, and three can't evolve smoothly into one! Well, actually it can. When space is filled with a condensate of quark pairs, the difference between three and one is negotiable.

Quark-hadron continuity is powerful, because it implies that the calculable forms of confinement and chiral symmetry breaking we construct by adapting the methods of BCS theory are in the same universality class as confinement and chiral symmetry breaking at low energies, within nuclear (or rather "nuclear") matter.

In real-world QCD we do not have three massless species of quarks, but two that are nearly massless (u, d) and one whose mass in certainly not negligible, at the chemical potentials relevant to nuclear matter. Nuclear matter at zero pressure is a very different beast from the CFL state, and does not seem amenable to similar theoretical analysis. Indeed, as mentioned previously, if we begin by ignoring the s quark, we find that candidate color superconducting states do not induce gaps for all the gluons and quarks, so we do not obtain a consistent starting point, free of infrared divergences. Although the CFL state is very likely favorable at asymptotically high densities, where the s quark mass becomes negligible, it is separated from nuclear matter by phase transitions, and one cannot draw continuous connections between their properties.

3. Explorations

3.1. *Exotic couplings of Nambu–Goldstone bosons*

Because the theme of symmetry supports many interesting variations, so does the theory of symmetry breaking. I will very briefly mention two that extend the spirit of our preceding discussion, and could become important:

- *Familons* are hypothetical Nambu–Goldsone bosons associated with the idea that the difference between particles in different families, which share most of their key properties (e.g., electron and muon) might arise by spontaneous symmetry breaking. They participate in flavor-changing transitions, and could be detectable even if their coupling is very weak.
- The concept of *Majorana mass* is usually considered in connection with spin-$\frac{1}{2}$ particles only. But the essential idea is more general, and simpler than it appears in that context. Consider adding a second scalar field ϕ_1 to our canonical model, with the symmetry

$$(\phi, \phi_1) \to e^{i\alpha}(\phi, \phi_1) \tag{18}$$

broken by $\langle \phi \rangle = v$ condensation. Mass terms arising from the symmetric interaction

$$\begin{aligned} \mathcal{L}_m &= -\kappa \phi^{2*} \phi_1^2 + \text{h.c.} \to -\kappa v^2 (\phi_1^2 + \phi_1^{*2}) \\ &= -2\kappa v^2 ((\text{Re } \phi_1)^2 - (\text{Im } \phi_1)^2) \end{aligned} \tag{19}$$

will split the quanta produced by the real and imaginary parts of ϕ_1, and thus tend to lift the degeneracy between quanta that had opposite $U(1)$ charge, and formed particle-antiparticle pairs, in the unbroken symmetry state. After symmetry breaking, they become two separate Majorana particles.

ϕ_1 could also have an intrinsic mass, unrelated to symmetry breaking. In general, the relative importance of characteristic "Majorana" effects, which mix particle and antiparticle, is non-universal, and can be large or small in different circumstances.

3.2. *Breaking τ (time translation) symmetry*

In 1964 Larkin, Ovchinnikov, Ferrell and Fulde (LOFF) proposed that possibility of pairing with electrons at a displaced momentum, i.e. electrons at momentum $\vec{k}$ with electrons at momentum $-\vec{k} + \delta$. In the 21st century, several probable realizations have been identified experimentally. Because momenta, in the quantum theory, are proportional to the wave vectors of electron wave functions, this sort of pairing involves non-trivial spatial periodicity of the order parameter. Thus the LOFF states provide a new sort of crystalline order.

It is natural to ask, whether one might have a similar phenomenon, involving ordering with non-trivial time dependence. To get oriented, I've found it very helpful to recall Anderson's pseudo-spin approach to BCS theory, which — after a simple generalization — supplies an illuminating model. Here I will sketch the essential idea.

In superconducting systems the absolute frequency dependence is rendered ambiguous by the possibility of time-dependent gauge transformations, or stated more simply by the lack of a natural zero of energy, so it is simpler to discuss particle-hole pairing.

For orientation purposes, let us begin by further specializing to the transparent case of two flat bands with energies $\varepsilon_1 < \varepsilon_2$, and the Hamiltonian

$$H = \frac{\varepsilon_2 + \varepsilon_1}{2} N + (\varepsilon_2 - \varepsilon_1)S_3 - g(S_-S_+ + S_+S_-)$$

$$= \frac{\varepsilon_2 + \varepsilon_1}{2} N + (\varepsilon_2 - \varepsilon_1)S_3 - 2g(\vec{S}^2 - S_3^2) \tag{20}$$

where

$$N = \sum_k b_k^\dagger b_k + \sum_k a_k^\dagger a_k \tag{21}$$

is the total occupation number and

$$S_+ = \sum_k b_k^\dagger a_k$$

$$= S_1 + iS_2$$

$$= S_-^\dagger$$

$$S_3 = \frac{1}{2}\left(\sum_k b_k^\dagger b_k - \sum_k a_k^\dagger a_k \right) \tag{22}$$

define hermitean pseudo-spin operators S_1, S_2, S_3 that satisfy the algebra of angular momentum and generate isospin-like rotations between the a and b modes.

Since $N, \vec{S}^2$, and S_3 commute, we can construct the minimum energy states for H, given N, by maximizing S (so $S = N/2$) and choosing a state with definite S_3. If S_3 is also allowed to vary, the minimum will occur for

$$\langle S_3 \rangle = \text{Max}\left(-\frac{\varepsilon_2 - \varepsilon_1}{4g}, -\frac{N}{2} \right). \tag{23}$$

(The second alternative on the right-hand side, which saturates the population of the a modes, is essentially trivial.) On the other hand it can be appropriate to hold the expectation value of S_3 fixed. We can imagine, for example, that the a and b modes correspond to states in distinct layers, whose total occupations can be fixed independently. (Note that the assumed interaction term does not require interlayer tunneling — this is just another way of saying that it commutes with S_3.) As we can see by re-arranging

$$a_k^\dagger b_k b_l^\dagger a_l \xrightarrow{\sim} -a_k^\dagger a_l b_l^\dagger b_k \qquad (24)$$

the assumed interaction corresponds, roughly, to an effective repulsion between density waves that does not depend on momentum transfer.

Now we can follow the classic BCS procedure, postulating a symmetry-breaking condensate. In this procedure, we assume the *ansatz*

$$\langle \mu, \theta | S_+ | \mu, \theta \rangle = \Delta_0 e^{i\theta} \qquad (25)$$

with Δ_0 a number, ultimately fixed self-consistently by the gap equation. Since

$$[H, S_+] = (\varepsilon_2 - \varepsilon_1) S_+ + 2g(S_3 S_+ + S_+ S_3) \qquad (26)$$

and

$$\frac{d}{dt}\langle S_+ \rangle = i\langle [H, S_+] \rangle \qquad (27)$$

consistent classical evolution for θ requires

$$\dot\theta = \varepsilon_2 - \varepsilon_1 + 4g\langle S_3 \rangle. \qquad (28)$$

This vanishes if $\langle S_3 \rangle$ is fixed non-trivially by Eq. (23), but not if that expectation value is pinned at a different value.

In the pseudo-spin formalism, this time dependence has a simple interpretation: The condensate is an effective spin of fixed magnitude at a fixed angle to the $\hat z$ axis, and the $\varepsilon_2 - \varepsilon_1$ term supplies an effective magnetic field in the $\hat z$ direction, which induces precession.

The BCS condensation *ansatz* is overkill for the flat-band model, where all states with the total spin and expectation value of S_3 are degenerate eigenstates. Its virtue is its ability, at the price of more complicated algebra, to accommodate more complex, momentum-dependent energies and interactions than assumed in Eq. (20). One expects qualitative aspects of spontaneous symmetry breaking to survive such generalizations. One can also consider bosonic systems along the same lines. Indeed, related techniques have been applied to discuss dynamic magnon condensation in liquid ^{3}He and density oscillations in two-component cold atomic gases. In both those contexts, very long-lived oscillatory states have been observed.

4. Reflections

By thinking hard and creatively about concrete problems of condensed matter physics, Phil Anderson achieved insights that have shaped, and continue to shape, our ideas about the elementary structure of Nature. He achieved those insights by identifying physical phenomena that present essentially new features, and capturing their conceptual essence. Their lessons could then be applied broadly, and orient us is wildly different physical regimes.

It is a beautiful thing, that inspires this birthday haiku:

More is Different.
Less is Different, too.
Same Difference.

Phil Anderson and Gauge Symmetry Breaking

Edward Witten

Institute for Advanced Study
Princeton, NJ 08540, USA
witten@ias.edu

In this article, I describe the celebrated paper that Phil Anderson wrote in 1962 with early contributions to the idea of gauge symmetry breaking in particle physics. To set the stage, I describe the work of Julian Schwinger to which Anderson was responding, and also some of Anderson's own work on superconductivity that provided part of the context. After describing Anderson's work I describe the later work of others, leading to the modern understanding of gauge symmetry breaking in weak interactions.

In this lecture, I am going to talk primarily about just one of the dozens of important papers that Phil Anderson has written. This paper was written just a little over 51 years ago, in 1962. I will set it in context by telling a little about the work that Julian Schwinger had done shortly before, and also we will recall Phil's earlier work on superconductivity that helped to set the stage. Then we will take a trip through later developments that occurred through the rest of the 1960s and beyond, and we will conclude with some observations about the present. (I won't talk about some early precursors such as Stueckelberg in 1938, but later we will get to the model introduced by Landau and Ginzburg in 1950.)

The title page of Phil's paper, which is called "Plasmons, Gauge Invariance, and Mass," and was received on November 8, 1962, can be found in Fig. 1. As one can see, Phil starts out by citing the work of Julian Schwinger. The reference is to two very short papers that Schwinger had written, also published in 1962. To understand Phil's work, we should first take a look at Schwinger's contributions.

In the first paper (Fig. 2), Schwinger argues somewhat abstractly that — in contrast to what we are familiar with in the case of electromagnetism — gauge invariance does not imply the existence of a massless spin-1 particle.

PHYSICAL REVIEW   VOLUME 130, NUMBER 1   1 APRIL 1963

Plasmons, Gauge Invariance, and Mass

P. W. ANDERSON

Bell Telephone Laboratories, Murray Hill, New Jersey

(Received 8 November 1962)

Schwinger has pointed out that the Yang-Mills vector boson implied by associating a generalized gauge transformation with a conservation law (of baryonic charge, for instance) does not necessarily have zero mass, if a certain criterion on the vacuum fluctuations of the generalized current is satisfied. We show that the theory of plasma oscillations is a simple nonrelativistic example exhibiting all of the features of Schwinger's idea. It is also shown that Schwinger's criterion that the vector field $m \neq 0$ implies that the matter spectrum before including the Yang-Mills interaction contains $m = 0$, but that the example of superconductivity illustrates that the physical spectrum need not. Some comments on the relationship between these ideas and the zero-mass difficulty in theories with broken symmetries are given.

R ECENTLY, Schwinger[1] has given an argument strongly suggesting that associating a gauge transformation with a local conservation law does not necessarily require the existence of a zero-mass vector boson. For instance, it had previously seemed impossible to describe the conservation of baryons in such a manner because of the absence of a zero-mass boson and of the accompanying long-range forces.[2] The problem of the mass of the bosons represents the major stumbling block in Sakurai's attempt to treat the dynamics of strongly interacting particles in terms of the Yang-Mills gauge fields which seem to be required to accompany the known conserved currents of baryon number and hypercharge.[3] (We use the term "Yang-Mills" in Sakurai's sense, to denote any generalized gauge field accompanying a local conservation law.)

The purpose of this article is to point out that the familiar plasmon theory of the free-electron gas exemplifies Schwinger's theory in a very straightforward manner. In the plasma, transverse electromagnetic waves do not propagate below the "plasma frequency," which is usually thought of as the frequency of longwavelength longitudinal oscillation of the electron gas. At and above this frequency, three modes exist, in close analogy (except for problems of Galilean invariance implied by the inequivalent dispersion of longitudinal and transverse modes) with the massive vector boson mentioned by Schwinger. The plasma frequency

is equivalent to the mass, while the finite density of electrons leading to divergent "vacuum" current fluctuations resembles the strong renormalized coupling of Schwinger's theory. In spite of the absence of low-frequency photons, gauge invariance and particle conservation are clearly satisfied in the plasma.

In fact, one can draw a direct parallel between the dielectric constant treatment of plasmon theory[4] and Schwinger's argument. Schwinger comments that the commutation relations for the gauge field A give us one sum rule for the vacuum fluctuations of A, while those for the matter field give a completely independent value for the fluctuations of matter current j. Since j is the source for A and the two are connected by field equations, the two sum rules are normally incompatible unless there is a contribution to the A rule from a free, homogeneous, weakly interacting, massless solution of the field equations. If, however, the source term is large enough, there can be no such contribution and the massless solutions cannot exist.

The usual theory of the plasmon does not treat the electromagnetic field quantum-mechanically or discuss vacuum fluctuations; yet there is a close relationship between the two arguments, and we, therefore, show that the quantum nature of the gauge field is irrelevant. Our argument is as follows:

The equation for the electromagnetic field is

$$p^2 A_\mu = (k^2 - \omega^2) A_\mu(\mathbf{k},\omega) = 4\pi j_\mu(\mathbf{k},\omega).$$

[1] J. Schwinger, Phys. Rev. **125**, 397 (1962).
[2] T. D. Lee and C. N. Yang, Phys. Rev. **98**, 1501 (1955).
[3] J. J. Sakurai, Ann. Phys. (N. Y.) **11**, 1 (1961).

[4] P. Nozières and D. Pines, Phys. Rev. **109**, 741 (1958).

Fig. 1. The first page of Phil Anderson's paper on gauge symmetry breaking, received November 8, 1962.

A couple of things are worth noting here, apart from the fact that Schwinger was forward-thinking to even ask the question. One is that Schwinger was motivated by the strong interactions (there is no mention of weak interactions in the paper). The question he asks in the first paragraph is whether the conservation law of baryon number could be a gauge symmetry. There is an obvious problem, which is that we do not see a massless spin-1 particle coupled to baryon number. So Schwinger asked whether it is possible for

PHYSICAL REVIEW   VOLUME 125, NUMBER 1   JANUARY 1, 1962

Gauge Invariance and Mass

JULIAN SCHWINGER

Harvard University, Cambridge, Massachusetts, and University of California, Los Angeles, California

(Received July 20, 1961)

It is argued that the gauge invariance of a vector field does not necessarily imply zero mass for an associated particle if the current vector coupling is sufficiently strong. This situation may permit a deeper understanding of nucleonic charge conservation as a manifestation of a gauge invariance, without the obvious conflict with experience that a massless particle entails.

DOES the requirement of gauge invariance for a vector field coupled to a dynamical current imply the existence of a corresponding particle with zero mass? Although the answer to this question is invariably given in the affirmative,[1] the author has become convinced that there is no such necessary implication, once the assumption of weak coupling is removed. Thus the path to an understanding of nucleonic (baryonic) charge conservation as an aspect of a gauge invariance, in strict analogy with electric charge,[2] may be open for the first time.

One potential source of error should be recognized at the outset. A gauge-invariant system is not the continuous limit of one that fails to admit such an arbitrary

Green's functions of other gauges have more complicated operator realizations, however, and will generally lack the positiveness properties of the radiation gauge.

Let us consider the simplest Green's function associated with the field $A_\mu(x)$, which can be derived from the unordered product

$$\langle A_\mu(x) A_\nu(x') \rangle$$

$$= \int \frac{(dp)}{(2\pi)^3} e^{ip(x-x')} dm^2 \, \eta_+(p) \delta(p^2+m^2) A_{\mu\nu}(p),$$

where the factor $\eta_+(p)\delta(p^2+m^2)$ enforces the spectral restriction to states with mass $m \geq 0$ and positive energy.

Fig. 2.   The first of Schwinger's two papers on gauge symmetry breaking.

baryon number (or something like baryon number) to be conserved because of a gauge symmetry, without the gauge symmetry producing a massless spin-1 particle.

The answer that Schwinger proposes — even in the first sentence of the abstract — is that gauge invariance does not necessarily imply the existence of a massless spin-1 particle if the coupling is large. His idea is that there is no massless particle if there is a pole in the current-current correlation function $\langle J_\mu(q) J_\nu(-q) \rangle$. The role of strong coupling is supposed to be to generate this pole. Thus, QED is weakly coupled, and has no such pole; Schwinger's point of view is that in a suitable gauge-invariant theory, strong coupling effects might produce such a pole.

In Schwinger's second paper (Fig. 3), he gives a concrete example of gauge invariance not implying the existence of a massless spin-1 particle. The example is based on a remarkable exact solution, not assuming weak coupling.

The model Schwinger solved was simply $(1+1)$-dimensional Quantum Electrodynamics, with electrons of zero bare mass. The action is

$$I = -\frac{1}{4e^2} \int d^2x \, F_{\mu\nu} F^{\mu\nu} + \int d^2x \, \bar{\Psi} i \slashed{D} \Psi.$$

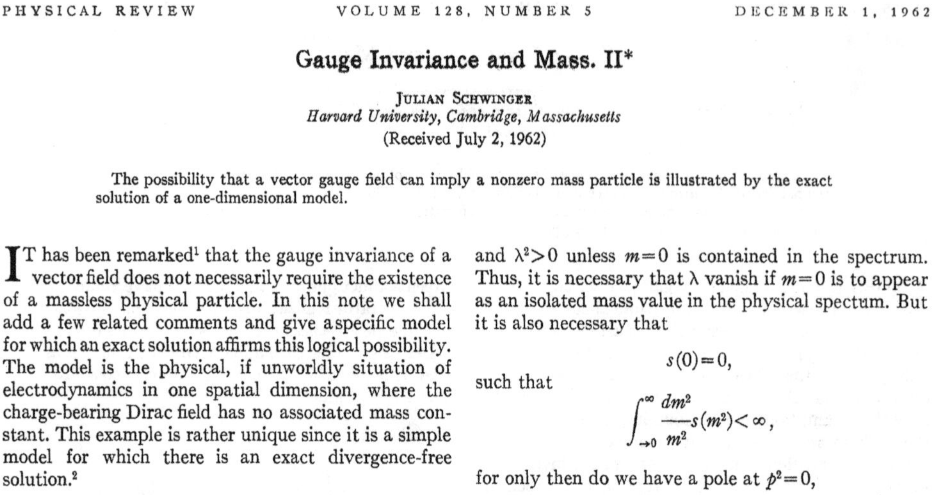

E. Witten

PHYSICAL REVIEW   VOLUME 128, NUMBER 5   DECEMBER 1, 1962

Gauge Invariance and Mass. II*

JULIAN SCHWINGER
Harvard University, Cambridge, Massachusetts
(Received July 2, 1962)

The possibility that a vector gauge field can imply a nonzero mass particle is illustrated by the exact solution of a one-dimensional model.

IT has been remarked[1] that the gauge invariance of a vector field does not necessarily require the existence of a massless physical particle. In this note we shall add a few related comments and give a specific model for which an exact solution affirms this logical possibility. The model is the physical, if unworldly situation of electrodynamics in one spatial dimension, where the charge-bearing Dirac field has no associated mass constant. This example is rather unique since it is a simple model for which there is an exact divergence-free solution.[2]

and $\lambda^2>0$ unless $m=0$ is contained in the spectrum. Thus, it is necessary that λ vanish if $m=0$ is to appear as an isolated mass value in the physical spectum. But it is also necessary that

such that

$$s(0)=0,$$

$$\int_{\to 0}^{\infty}\frac{dm^2}{m^2}s(m^2)<\infty,$$

for only then do we have a pole at $p^2=0$,

Fig. 3. Schwinger's second paper on gauge symmetry breaking, in which he introduced and solved what is now called the Schwinger model.

Nowadays this model — which is known as the Schwinger model — is usually solved simply and understandably (but surprisingly) by "bosonization," which converts it to a free theory (of the gauge field A_μ and a scalar field ϕ) with all the properties that Schwinger claimed. Schwinger's approach to solving it was more axiomatic.

The model is actually considered an important example that illustrates quite a few things, and not only what Schwinger had in mind. For instance, it is also used to illustrate the physics of a gauge theory θ-angle, and after a small perturbation to give the electron a bare mass, it becomes a model of confinement of charged particles. However, although I was not in physics at the time, I suspect that Schwinger's extremely short paper mystified many of his contemporaries. His way of solving the model probably was a little abstract, and the whole thing probably seemed to revolve around peculiarities of $1+1$ dimensions.

Schwinger's concept was summarized in the last sentence of his first paper: "the essential point is embodied in the view that the observed physical world is the outcome of the dynamical play among underlying primary fields, and the relationship between these fundamental fields and the phenomenological particles can be comparatively remote, in contrast to the immediate correlation that is commonly assumed." In other words, in general, there need be no simple relationship between particles and fields — or in condensed matter physics, between bare electrons and nuclei and the emergent quasi-particles that give a more useful description at long distances.

Schwinger is saying that the situation that prevails in QED — in which the electron field corresponds to electrons, and the photon field corresponds to photons — results from the fact that this theory is weakly coupled. In a strongly coupled theory, there might be no simple correspondence between fields and particles. This was actually a very wise remark, probably putting Schwinger way ahead of his contemporaries in particle physics. And it is at the core of the way we now understand the strong interactions.

But Phil Anderson showed that Schwinger was actually not entirely correct about the specific question he was writing about — how to have gauge invariance without a massless spin-1 particle. To be more precise, everything that Schwinger said about strong coupling is true, but it is not the whole story. As Anderson showed, a weakly coupled vector meson might also acquire a mass, and here the essence of the matter is not strong coupling but symmetry breaking — the properties of the vacuum. In fact, Phil expressed a point of view that is quite opposite to Schwinger's, showing that not just strong coupling but even quantum mechanics is irrelevant to the problem of how to have gauge invariance without a massless spin-1 particle.

Phil's paper is largely devoted to two examples from well-established physics. He begins by saying that "the familiar plasmon theory of the free electron gas exemplifies Schwinger's theory in a very straightforward manner," with the plasma frequency, below which electromagnetic waves do not propagate, playing the role of the vector meson mass for Schwinger. He shows that the usual analysis of screening in a plasma can be put in close parallel with what Schwinger had said in the relativistic case. He also observes that the problem of screening in a plasma is usually understood classically, without invoking quantum mechanics, and deduces that "the quantum nature of the gauge field is irrelevant" to the question of how to have gauge invariance without a massless vector particle.

The second part of Phil's 1962 paper deals with an example that is even more incisive — superconductivity. The background to this was provided by a series of three papers that Phil had written in 1958 (shown in Fig. 4 is the title page of the first of the three papers — which incidentally is the one he referred to in 1962). In these papers, Phil had analyzed gauge invariance and the fate of the "Goldstone" boson (the term is ahistorical as Goldstone had not yet formulated his relativistic theorem) in the BCS theory of superconductivity, showing that this mode combines with ordinary photons to become a gapped state of spin-1. Thus the electronic state of an ordinary BCS superconductor is truly gapless.

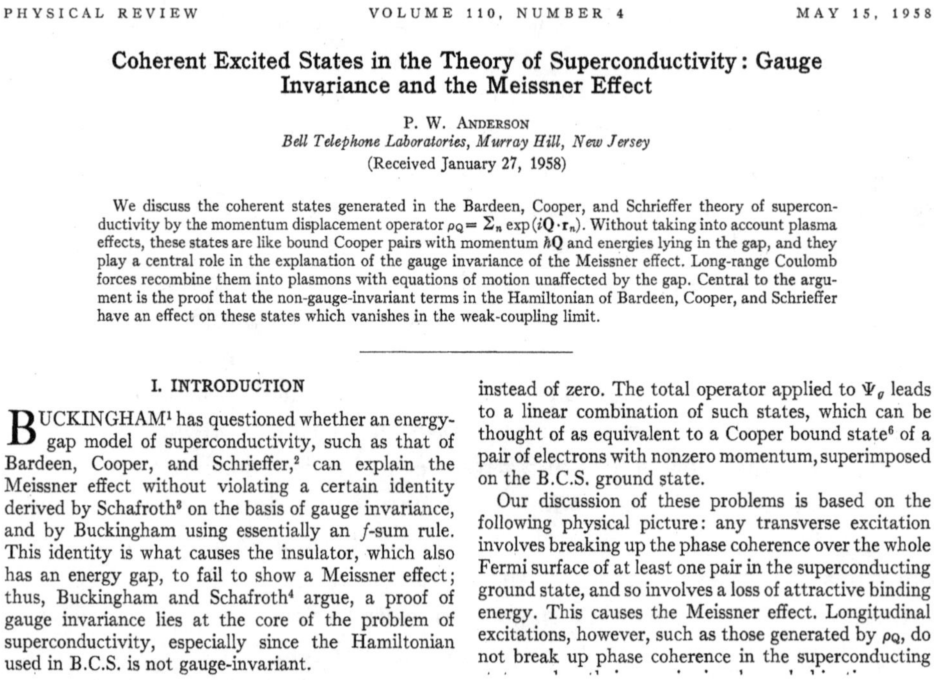

PHYSICAL REVIEW VOLUME 110, NUMBER 4 MAY 15, 1958

Coherent Excited States in the Theory of Superconductivity: Gauge Invariance and the Meissner Effect

P. W. ANDERSON
Bell Telephone Laboratories, Murray Hill, New Jersey
(Received January 27, 1958)

We discuss the coherent states generated in the Bardeen, Cooper, and Schrieffer theory of superconductivity by the momentum displacement operator $\rho_Q = \Sigma_n \exp(i\mathbf{Q} \cdot \mathbf{r}_n)$. Without taking into account plasma effects, these states are like bound Cooper pairs with momentum $\hbar\mathbf{Q}$ and energies lying in the gap, and they play a central role in the explanation of the gauge invariance of the Meissner effect. Long-range Coulomb forces recombine them into plasmons with equations of motion unaffected by the gap. Central to the argument is the proof that the non-gauge-invariant terms in the Hamiltonian of Bardeen, Cooper, and Schrieffer have an effect on these states which vanishes in the weak-coupling limit.

I. INTRODUCTION

BUCKINGHAM[1] has questioned whether an energy-gap model of superconductivity, such as that of Bardeen, Cooper, and Schrieffer,[2] can explain the Meissner effect without violating a certain identity derived by Schafroth[3] on the basis of gauge invariance, and by Buckingham using essentially an f-sum rule. This identity is what causes the insulator, which also has an energy gap, to fail to show a Meissner effect; thus, Buckingham and Schafroth[4] argue, a proof of gauge invariance lies at the core of the problem of superconductivity, especially since the Hamiltonian used in B.C.S. is not gauge-invariant.

instead of zero. The total operator applied to Ψ_g leads to a linear combination of such states, which can be thought of as equivalent to a Cooper bound state[6] of a pair of electrons with nonzero momentum, superimposed on the B.C.S. ground state.

Our discussion of these problems is based on the following physical picture: any transverse excitation involves breaking up the phase coherence over the whole Fermi surface of at least one pair in the superconducting ground state, and so involves a loss of attractive binding energy. This causes the Meissner effect. Longitudinal excitations, however, such as those generated by ρ_Q, do not break up phase coherence in the superconducting

Fig. 4. The first of Anderson's three papers of 1958 on gauge invariance in the BCS model of superconductivity.

In the 1962 paper, Phil explains cogently how superconductivity illustrates the phenomenon described by Schwinger, in a context in which the gauge field is weakly coupled and the physics is well understood. Much of this paper reads just like what one would explain to a student today. For example: "the way is now open for a degenerate vacuum theory of the Nambu type without any difficulties involving either zero-mass Yang–Mills gauge bosons or zero-mass Goldstone bosons. These two types of bosons seem capable of 'canceling each other out' and leaving finite mass bosons only." He goes on: "It is not at all clear that the way for a Sakurai theory [with baryon number as a gauge symmetry] is equally uncluttered. The only mechanism suggested by the present work (of course, we have not discussed non-abelian gauge groups) for giving the gauge field mass is the degenerate vacuum type of theory, in which the original symmetry is not manifest in the observable domain. Therefore, it needs to be demonstrated that the necessary conservation laws can be maintained." In other words, as one would say today, if baryon number is gauged and spontaneously broken, then baryon number will not be conserved in nature.

PHYSICAL REVIEW   VOLUME 130, NUMBER 1   1 APRIL 1963

P. W. ANDERSON

I should like to close with one final remark on the Goldstone theorem. This theorem was initially conjectured, one presumes, because of the solid-state analogs, via the work of Nambu[10] and of Anderson.[11] The theorem states, essentially, that if the Lagrangian possesses a continuous symmetry group under which the ground or vacuum state is not invariant, that state is, therefore, degenerate with other ground states. This implies a zero-mass boson. Thus, the solid crystal violates translational and rotational invariance, and possesses phonons; liquid helium violates (in a certain sense only, of course) gauge invariance, and possesses a longitudinal phonon; ferro-magnetism violates spin rotation symmetry, and possesses spin waves; superconductivity violates gauge invariance, and would have a zero-mass collective mode in the absence of long-range Coulomb forces.

It is noteworthy that in most of these cases, upon closer examination, the Goldstone bosons do indeed become tangled up with Yang-Mills gauge bosons and, thus, do not in any true sense really have zero mass. Superconductivity is a familiar example, but a similar phenomenon happens with phonons; when the phonon frequency is as low as the gravitational plasma frequency, $(4\pi G\rho)^{1/2}$ (wavelength $\sim 10^4$ km in normal matter) there is a phonon-graviton interaction: in that case, because of the peculiar sign of the gravitational interaction, leading to instability rather than finite mass.[12] Utiyama[13] and Feynman have pointed out that gravity is also a Yang-Mills field. It is an amusing observation that the three phonons plus two gravitons are just enough components to make up the appropriate tensor particle which would be required for a finite-mass graviton.

Spin waves also are known to interact strongly with magnetostatic forces at very long wavelengths,[14] for rather more obscure and less satisfactory reasons. We conclude, then, that the Goldstone zero-mass difficulty is not a serious one, because we can probably cancel it off against an equal Yang-Mills zero-mass problem. What is not clear yet, on the other hand, is whether it is possible to describe a truly strong conservation law such as that of baryons with a gauge group and a Yang-Mills field having finite mass.

I should like to thank Dr. John R. Klauder for valuable conversations and, particularly, for correcting some serious misapprehensions on my part, and Dr. John G. Taylor for calling my attention to Schwinger's work.

---

[12] J. H. Jeans, Phil. Trans. Roy. Soc. London 101, 157 (1903).
[13] R. Utiyama, Phys. Rev. 101, 1597 (1956); R. P. Feynman (unpublished).
[14] L. R. Walker, Phys. Rev. 105, 390 (1957).

Fig. 5.   The concluding part of Anderson's 1962 paper on gauge symmetry breaking.

And let us look at the end of the paper (Fig. 5): "I should like to close with one final remark on the Goldstone theorem. This theorem was initially conjectured, one presumes, because of the solid-state analogs, via the work of Nambu and Anderson" (the reference here is to Nambu's work on spontaneously broken chiral symmetry and to Anderson's paper whose title page appears in Fig. 4). He goes on to give various examples, both old (spin waves and phonons) and new (superconductors and superfluids). And then he writes, "It is noteworthy that in most of these cases, upon closer examination, the Goldstone bosons do indeed become tangled up with Yang–Mills gauge bosons and do not in any true sense really have zero mass. Superconductivity is a familiar example, but a similar phenomenon happens with phonons; when the phonon frequency is as low as the gravitational plasma frequency, $(4\pi G)^{1/2}$ (wavelength $\sim 10^4$ km in normal matter) there is a phonon-graviton interaction: in that case, because of the peculiar sign of the gravitational interaction, leading to instability rather than finite mass. Utiyama and Feynman have pointed out that gravity is also a Yang–Mills field. It is an amusing observation that three phonons plus two gravitons are just enough components to make up the appropriate tensor particle which would be required for a finite-mass graviton." So the answer to the question of who first tried to discuss a gravitational analog of gauge symmetry breaking is that Anderson did.

What happened next? In 1964, Peter Higgs wrote two papers on gauge invariance with massive vector particles in relativistic physics. Like Anderson, but unlike Schwinger, his starting point is spontaneous breaking of symmetry. This was a few years after Goldstone's theorem and particle physicists were more familiar with this concept. In the first paper, he explains somewhat abstractly, in a language similar to Schwinger's, why Goldstone's theorem is not valid in the case of a gauge symmetry. Higgs's paper that really had greater impact was the second one in which he described a concrete (and weakly coupled) model that everyone could understand and also introduced the Higgs particle. Let us take a look at this paper (Fig. 6).

Higgs explains at the outset that the phenomenon of a gauge boson acquiring a mass via symmetry breaking "is just the relativistic analog of the plasmon phenomenon to which Anderson has drawn attention: that the scalar zero-mass excitations of a superconducting neutral Fermi gas become longitudinal plasmon modes of finite mass when the gas is charged." He then goes on, in the next paragraph (bottom left in Fig. 6), to write down his model.

Higgs's model was simply a relativistic version of the model that Landau and Ginzburg had introduced to describe superconductivity. (Neither Higgs nor any of the authors I have mentioned cited Landau and Ginzburg. Higgs describes his model as a gauge-invariant version of a model of Goldstone.) The Landau–Ginzburg model can be deduced from the action

$$I = \int d^3x \, dt \left(-\frac{1}{4} F_{\mu\nu} F^{\mu\nu} + i\bar{\Phi} \frac{D}{Dt} \Phi - \frac{1}{2m} \sum_i \left| \frac{D\Phi}{Dx^i} \right|^2 - \lambda(\bar{\Phi}\Phi - a^2)^2 \right).$$

Higgs's model, which particle physicists call the abelian Higgs model, is the same thing (apart from minor rescalings) with the kinetic energy of the scalar field made relativistic

$$I = \int d^3x \, dt \left(-\frac{1}{4} F_{\mu\nu} F^{\mu\nu} + D_\mu \bar{\Phi} D^\mu \Phi - \lambda(\bar{\Phi}\Phi - a^2)^2 \right).$$

The models are the same except that the abelian Higgs model is quadratic rather than linear in time derivatives.

When Dirac (for spin-$\frac{1}{2}$) or Klein and Gordon (for spin-0) made the Schrödinger equation relativistic, they introduced an extra degree of freedom (the antiparticle). Something somewhat similar happens in making the

VOLUME 13, NUMBER 16   PHYSICAL REVIEW LETTERS   19 OCTOBER 1964

BROKEN SYMMETRIES AND THE MASSES OF GAUGE BOSONS

Peter W. Higgs
Tait Institute of Mathematical Physics, University of Edinburgh, Edinburgh, Scotland
(Received 31 August 1964)

In a recent note[1] it was shown that the Goldstone theorem,[2] that Lorentz-covariant field theories in which spontaneous breakdown of symmetry under an internal Lie group occurs contain zero-mass particles, fails if and only if the conserved currents associated with the internal group are coupled to gauge fields. The purpose of the present note is to report that, as a consequence of this coupling, the spin-one quanta of some of the gauge fields acquire mass; the longitudinal degrees of freedom of these particles (which would be absent if their mass were zero) go over into the Goldstone bosons when the coupling tends to zero. This phenomenon is just the relativistic analog of the plasmon phenomenon to which Anderson[3] has drawn attention: that the scalar zero-mass excitations of a superconducting neutral Fermi gas become longitudinal plasmon modes of finite mass when the gas is charged.

The simplest theory which exhibits this behavior is a gauge-invariant version of a model used by Goldstone[2] himself: Two real[4] scalar fields φ_1, φ_2 and a real vector field A_μ interact through the Lagrangian density

$$L = -\tfrac{1}{2}(\nabla \varphi_1)^2 - \tfrac{1}{2}(\nabla \varphi_2)^2$$
$$- V(\varphi_1{}^2 + \varphi_2{}^2) - \tfrac{1}{4} F_{\mu\nu} F^{\mu\nu}, \quad (1)$$

about the "vacuum" solution $\varphi_1(x) = 0$, $\varphi_2(x) = \varphi_0$:

$$\partial^\mu \{\partial_\mu(\Delta\varphi_1) - e\varphi_0 A_\mu\} = 0, \quad (2a)$$

$$\{\partial^2 - 4\varphi_0{}^2 V''(\varphi_0{}^2)\}(\Delta\varphi_2) = 0, \quad (2b)$$

$$\partial_\nu F^{\mu\nu} = e\varphi_0\{\partial^\mu(\Delta\varphi_1) - e\varphi_0 A^\mu\}. \quad (2c)$$

Equation (2b) describes waves whose quanta have (bare) mass $2\varphi_0\{V''(\varphi_0{}^2)\}^{1/2}$; Eqs. (2a) and (2c) may be transformed, by the introduction of new variables

$$B_\mu = A_\mu - (e\varphi_0)^{-1}\partial_\mu(\Delta\varphi_1),$$

$$G_{\mu\nu} = \partial_\mu B_\nu - \partial_\nu B_\mu = F_{\mu\nu}, \quad (3)$$

into the form

$$\partial_\mu B^\mu = 0, \quad \partial_\nu G^{\mu\nu} + e^2\varphi_0{}^2 B^\mu = 0. \quad (4)$$

Equation (4) describes vector waves whose quanta have (bare) mass $e\varphi_0$. In the absence of the gauge field coupling ($e = 0$) the situation is quite different: Equations (2a) and (2c) describe zero-mass scalar and vector bosons, respectively. In passing, we note that the right-hand side of (2c) is just the linear approximation to the conserved current: It is linear in the vector potential,

Fig. 6. Peter Higgs's second paper on gauge symmetry breaking, in which he introduced in Eq. (1) what particle physicists know as the abelian Higgs model.

Landau–Ginzburg model relativistic. There is an extra degree of freedom because Φ and $\bar{\Phi}$ become independent rather than being canonically conjugate, as they are in Landau–Ginzburg theory. In Landau–Ginzburg theory, if we ignore the gauge fields, there is one spin-0 particle, the Goldstone boson, but if we include the gauge fields, then — as Anderson had explained in the more sophisticated context of BCS theory — it becomes part of a massive spin-1 particle. In the abelian Higgs model, there is a second and massive spin-0 mode — this is the fluctuation in the magnitude of Φ, which is now usually called the Higgs particle.

Actually, although there is not quite a Higgs particle in usual models of superconductivity — or in superconducting phenomenology — there is a close cousin. In a superconductor, there are two characteristic lengths, the penetration depth and the coherence length. They are described in the Landau–Ginzburg and BCS models of superconductivity and are measured experimentally. (The difference between a Type I and Type II superconductor has to do with which is bigger.) These are the analogs of the gauge boson mass and the Higgs boson mass in particle physics. Relativistically, the rate at which the field decays in space is related to a particle mass, but non-relativistically there is no reason for this to happen, and in the Landau–Ginzburg model it doesn't, in the case of the correlation length. The Landau–Ginzburg model and the abelian Higgs model are completely equivalent for static phenomena since they coincide once one drops the time derivatives.

Similar ideas were developed by others at roughly the same time as Higgs. We will just take a quick look. The paper of Englert and Brout is in Fig. 7. This paper is notable for considering symmetry breaking in non-abelian gauge theory, while previous authors had considered the abelian case, sometimes saying that this was for simplicity. "The importance of this problem," they say, "lies in the possibility that strong-interaction physics originates from massive gauge fields coupled to a system of conserved currents," for which they refer to Sakurai. Soon after was the paper of Guralnik, Hagen, and Kibble (Fig. 8), followed by Migdal and Polyakov (Fig. 9). The title of Migdal and Polyakov, "Spontaneous Breakdown of Strong Interaction Symmetry and Absence of Massless Particles," shows that they, too, were

VOLUME 13, NUMBER 9   PHYSICAL REVIEW LETTERS   31 AUGUST 1964

BROKEN SYMMETRY AND THE MASS OF GAUGE VECTOR MESONS*

F. Englert and R. Brout
Faculté des Sciences, Université Libre de Bruxelles, Bruxelles, Belgium
(Received 26 June 1964)

It is of interest to inquire whether gauge vector mesons acquire mass through interaction[1]; by a gauge vector meson we mean a Yang-Mills field[2] associated with the extension of a Lie group from global to local symmetry. The importance of this problem resides in the possibility that strong-interaction physics originates from massive gauge fields related to a system of conserved currents.[3] In this note, we shall show that in certain cases vector mesons do indeed acquire mass when the vacuum is degenerate with respect to a compact Lie group.

those vector mesons which are coupled to currents that "rotate" the original vacuum are the ones which acquire mass [see Eq. (6)].

We shall then examine a particular model based on chirality invariance which may have a more fundamental significance. Here we begin with a chirality-invariant Lagrangian and introduce both vector and pseudovector gauge fields, thereby guaranteeing invariance under both local phase and local γ_5-phase transformations. In this model the gauge fields themselves may break the γ_5 invariance leading to a mass for the original Fermi field. We shall show in this case

Fig. 7.   Brout and Englert on gauge symmetry breaking.

VOLUME 13, NUMBER 20   PHYSICAL REVIEW LETTERS   16 NOVEMBER 1964

GLOBAL CONSERVATION LAWS AND MASSLESS PARTICLES*

G. S. Guralnik,[†] C. R. Hagen,[‡] and T. W. B. Kibble

Department of Physics, Imperial College, London, England

(Received 12 October 1964)

In all of the fairly numerous attempts to date to formulate a consistent field theory possessing a broken symmetry, Goldstone's remarkable theorem[1] has played an important role. This theorem, briefly stated, asserts that if there exists a conserved operator Q_i such that

$$[Q_i, A_j(x)] = \sum_k t_{ijk} A_k(x),$$

and if it is possible consistently to take $\sum_k t_{ijk} \times \langle 0|A_k|0\rangle \neq 0$, then $A_j(x)$ has a zero-mass particle in its spectrum. It has more recently been observed that the assumed Lorentz invariance essential to the proof[2] may allow one the hope of avoiding such massless particles through the in-troduction of vector gauge fields and the consequent breakdown of manifest covariance.[3] This, of course, represents a departure from the assumptions of the theorem, and a limitation on its applicability which in no way reflects on the general validity of the proof.

In this note we shall show, within the framework of a simple soluble field theory, that it is possible consistently to break a symmetry (in the sense that $\sum_k t_{ijk}\langle 0|A_k|0\rangle \neq 0$) without requiring that $A(x)$ excite a zero-mass particle. While this result might suggest a general procedure for the elimination of unwanted massless bosons, it will be seen that this has been accomplished by giving up the global conservation law usually

585

Fig. 8. Guralnik, Hagen, and Kibble on gauge symmetry breaking.

SOVIET PHYSICS JETP   VOLUME 24, NUMBER 1   JANUARY, 1967

SPONTANEOUS BREAKDOWN OF STRONG INTERACTION SYMMETRY AND THE ABSENCE OF MASSLESS PARTICLES

A. A. MIGDAL and A. M. POLYAKOV

Submitted to JETP editor November 30, 1965; resubmitted February 16, 1966

J. Exptl. Theoret. Physics (U.S.S.R.) **51**, 135−146 (July, 1966)

The occurrence of massless particles in the presence of spontaneous symmetry breakdown is discussed. By summing all Feynman diagrams, one obtains for the difference of the mass operators $M_a(p) - M_b(p)$ of particles a and b belonging to a supermultiplet an equation which is identical to the Bethe–Salpeter equation for the wave function of a scalar bound state of vanishing mass (a "zeron") in the annihilation channel ab of the corresponding particles. It is shown that if symmetry is spontaneously violated in a Yang–Mills type theory involving vector mesons, the zerons interact only with virtual particles and therefore unobservable. On the other hand, the vector mesons acquire a mass in spite of the generalized gauge invariance. It is shown in Appendices A and B that the asymmetrical solution corresponds to a minimal energy of the vacuum and that C-invariance of the solution implies strangeness conservation for it.

1. INTRODUCTION

SPONTANEOUS symmetry breakdown related to an instability of the system under consideration with respect to an infinitesimally weak asymmetric perturbation is often encountered in quantum statistical mechanics (ferromagnetism, superconductivity, etc.). A large number of such examples has been analyzed in Bogolyubov's review.[1]

Fig. 9. Migdal and Polyakov on gauge symmetry breaking.

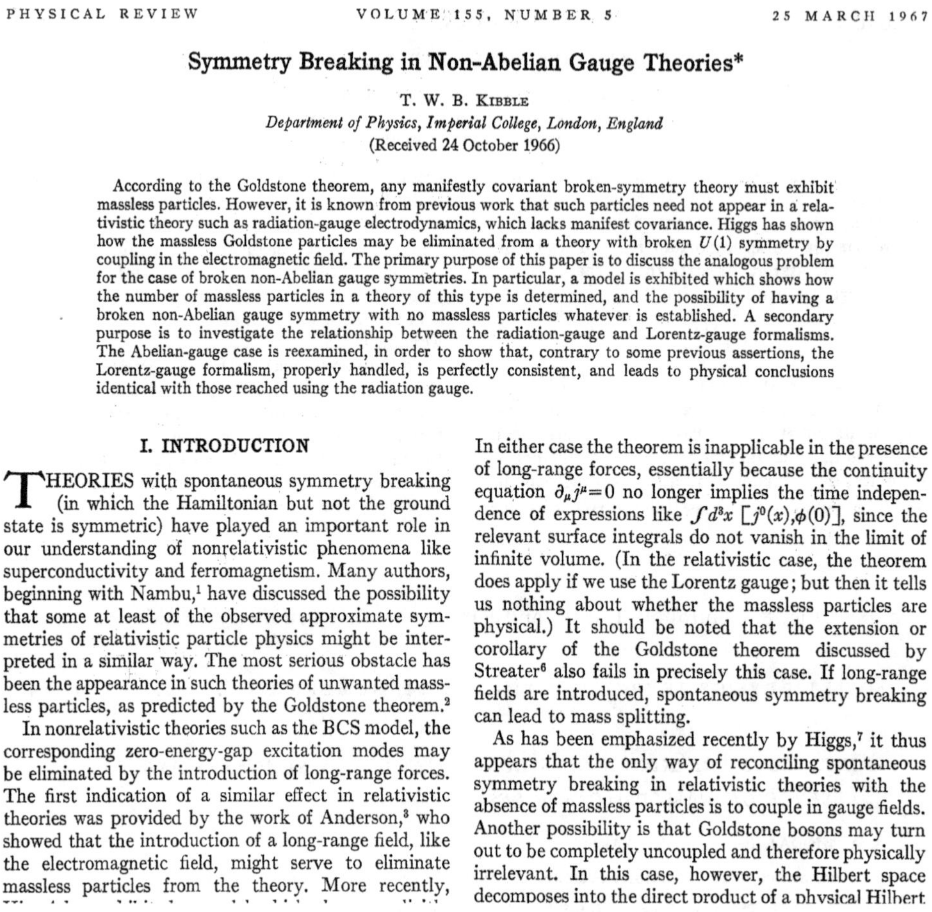

PHYSICAL REVIEW   VOLUME 155, NUMBER 5   25 MARCH 1967

Symmetry Breaking in Non–Abelian Gauge Theories*

T. W. B. KIBBLE

Department of Physics, Imperial College, London, England

(Received 24 October 1966)

According to the Goldstone theorem, any manifestly covariant broken-symmetry theory must exhibit massless particles. However, it is known from previous work that such particles need not appear in a relativistic theory such as radiation-gauge electrodynamics, which lacks manifest covariance. Higgs has shown how the massless Goldstone particles may be eliminated from a theory with broken $U(1)$ symmetry by coupling in the electromagnetic field. The primary purpose of this paper is to discuss the analogous problem for the case of broken non-Abelian gauge symmetries. In particular, a model is exhibited which shows how the number of massless particles in a theory of this type is determined, and the possibility of having a broken non-Abelian gauge symmetry with no massless particles whatever is established. A secondary purpose is to investigate the relationship between the radiation-gauge and Lorentz-gauge formalisms. The Abelian-gauge case is reexamined, in order to show that, contrary to some previous assertions, the Lorentz-gauge formalism, properly handled, is perfectly consistent, and leads to physical conclusions identical with those reached using the radiation gauge.

I. INTRODUCTION

THEORIES with spontaneous symmetry breaking (in which the Hamiltonian but not the ground state is symmetric) have played an important role in our understanding of nonrelativistic phenomena like superconductivity and ferromagnetism. Many authors, beginning with Nambu,[1] have discussed the possibility that some at least of the observed approximate symmetries of relativistic particle physics might be interpreted in a similar way. The most serious obstacle has been the appearance in such theories of unwanted massless particles, as predicted by the Goldstone theorem.[2]

In nonrelativistic theories such as the BCS model, the corresponding zero-energy-gap excitation modes may be eliminated by the introduction of long-range forces. The first indication of a similar effect in relativistic theories was provided by the work of Anderson,[3] who showed that the introduction of a long-range field, like the electromagnetic field, might serve to eliminate massless particles from the theory. More recently,

In either case the theorem is inapplicable in the presence of long-range forces, essentially because the continuity equation $\partial_\mu j^\mu = 0$ no longer implies the time independence of expressions like $\int d^3x \, [j^0(x), \phi(0)]$, since the relevant surface integrals do not vanish in the limit of infinite volume. (In the relativistic case, the theorem does apply if we use the Lorentz gauge; but then it tells us nothing about whether the massless particles are physical.) It should be noted that the extension or corollary of the Goldstone theorem discussed by Streater[6] also fails in precisely this case. If long-range fields are introduced, spontaneous symmetry breaking can lead to mass splitting.

As has been emphasized recently by Higgs,[7] it thus appears that the only way of reconciling spontaneous symmetry breaking in relativistic theories with the absence of massless particles is to couple in gauge fields. Another possibility is that Goldstone bosons may turn out to be completely uncoupled and therefore physically irrelevant. In this case, however, the Hilbert space decomposes into the direct product of a physical Hilbert

Fig. 10. Kibble on symmetry breaking in non-abelian gauge theory.

thinking of the strong interactions as the arena in which gauge symmetry breaking might play a role.

From here, let us move forward to Kibble in 1966 (Fig. 10). After mentioning the example of superconductivity, Kibble writes "The first indication of a similar effect in relativistic theories was provided by the work of Anderson, who showed that the introduction of a long-range field, like the electromagnetic field, might serve to eliminate massless particles from the theory. More recently, Higgs has exhibited a model which shows explicitly how the massless Goldstone bosons are eliminated by coupling the current associated with the broken symmetry to a gauge field." He then goes on to discuss some important details of symmetry breaking in non-abelian gauge theory. He explains how it it is possible to have partial breaking of non-abelian gauge symmetry, with some gauge mesons remaining massless. Like Higgs and some of the others, he does not really say what the physical application is supposed to be, but he does remark that nature has only one massless vector particle — the photon — but various (in some cases approximate) global symmetries. At least this was on the right track.

The next milestone, of course, was that in 1967–8, Weinberg and Salam actually found what spontaneous gauge symmetry breaking is good for in particle physics. (Their model was a gauge-invariant refinement of an earlier model by Glashow. That model had W and Z mesons, but lacked the relationship between their masses and couplings that follows from the spontaneous symmetry breaking mechanism introduced by Weinberg and Salam.) However, since we have already looked at quite a few original papers, let us jump ahead to Weinberg's Nobel Prize address in 1979.

In the passage copied in Fig. 11, Weinberg explains quite vividly how — like everyone else in the 1960s, it seems — he started by assuming that gauge symmetry breaking was supposed to be applied to the strong interactions. His detailed explanation actually makes interesting reading. To help the reader understand this passage, I will make the following remarks. If baryon number is a gauge symmetry, what is the gauge meson? The lightest hadronic particle of spin-1 with the appropriate quantum numbers is the ω meson, or the ϕ meson if one includes strange particles. So one might think of one of those as a gauge meson. But if baryon number is a gauge symmetry, perhaps isospin symmetry is a gauge symmetry also. In this case, the lightest candidates for the massive gauge particles are the ρ mesons. But bearing in mind that isospin symmetry is part of a spontaneously broken $SU(2) \times SU(2)$ chiral symmetry, perhaps there is also an axial vector triplet of massive gauge mesons; the A_1 is the lightest candidate. All this is quite alien to present-day thinking, and as Weinberg explains, there were a lot of problems: massless ρ mesons, or no pions, or explicit (rather than spontaneous) breaking of gauge invariance and therefore no renormalizability, depending on what assumptions he made.

Then enlightenment dawns. Weinberg explains that "At some point in the fall of 1967, I think while driving to my office at M.I.T., it occurred to me that I had been applying the right ideas to the wrong problem. It is not the ρ mesons that is massless: it is the photon. And its partner is not the A_1, but the massive intermediate boson, which since the time of Yukawa had been suspected to be the mediator of the weak interactions. The weak and electromagnetic interactions could then be described in a unified way in terms of an exact but spontaneously broken gauge symmetry.... And this

CONCEPTUAL FOUNDATIONS OF THE UNIFIED THEORY OF WEAK AND ELECTROMAGNETIC INTERACTIONS
Nobel Lecture, December 8, 1979
by STEVEN WEINBERG
Lyman Laboratory of Physics Harvard University and Harvard-Smithsonian Center for Astrophysics, Cambridge, MASS., USA.

Now, back to 1967. I had been considering the implications of the broken SU(2) x SU(2) symmetry of the strong interactions, and I thought of trying out the idea that perhaps the SU(2) x SU(2) symmetry was a "local," not merely a "global," symmetry. That is, the strong interactions might be described by something like a Yang-Mills theory, but in addition to the vector ϱ mesons of the Yang-Mills theory, there would also be axial vector Al mesons. To give the ϱ meson a mass, it was necessary to insert a common ϱ and Al mass term in the Lagrangian, and the spontaneous

breakdown of the SU(2) x SU(2) symmetry would then split the ϱ and Al by something like the Higgs mechanism, but since the theory would not be gauge invariant the pions would remain as physical Goldstone bosons. This theory gave an intriguing result, that the Al/ϱ mass ratio should be $\sqrt{2}$, and in trying to understand this result without relying on perturbation theory, I discovered certain sum rules, the "spectral function sum rules," [23] which turned out to have variety of other uses. But the SU(2) x SU(2) theory was not gauge invariant, and hence it could not be renormalizable, [24] so I was not too enthusiastic about it. [25] Of course, if I did not insert the ϱ-Al mass term in the Lagrangian, then the theory would be gauge invariant and renormalizable, and the Al would be massive. But then there would be no pions and the ϱ mesons would be massless, in obvious contradiction (to say the least) with observation.

Fig. 11.   A passage from Steve Weinberg's Nobel Prize Lecture in 1979.

theory would be renormalizable like quantum electrodynamics because it is gauge invariant like quantum electrodynamics."

I have been asked whether Weinberg and Salam were the first to use gauge symmetry breaking to give masses to particles other than gauge bosons. They were the first to generate masses for leptons in this way. For strong interactions, matters are more complicated. The modern understanding is that hadron masses come partly from dynamical effects of QCD and partly from the bare masses of quarks and leptons. (The masses of protons, neutrons, and pions come mostly from the QCD effects while heavier hadrons containing charm or bottom quarks get mass mostly from the quark bare masses.) In the modern understanding, it is the quark bare masses, not the part of the hadron masses coming from QCD effects, that result from gauge symmetry breaking and the coupling to the Higgs particle. Such a clear statement was however not possible until QCD was put in its modern form in 1973, enabling the full formulation of the Standard Model. From a modern point of view, earlier attempts to connect hadron masses to gauge symmetry breaking (as opposed to spontaneous breaking of global chiral symmetries) mostly did not focus on the right part of the problem.

The emergence of the Standard Model brings us to the modern era. I will conclude this talk by sketching briefly a few of the subsequent developments. First we will talk about the strong interactions. Since the discovery of asymptotic freedom in 1973, we describe the strong interactions via an *unbroken* non-abelian gauge theory with gauge group $SU(3)$, coupled to quarks. The $SU(3)$ gauge symmetry is definitely unbroken, so at first sight it looks like spontaneous breaking of gauge symmetry turned out to be the wrong idea for the strong interactions.

There is a mystery, however, in QCD: why don't we see the quarks? Experiment and computer simulations both seem to show that the quarks are "confined," that the energy grows indefinitely if one tries to separate a quark from an antiquark. Confinement is quite a surprise and I would say that we still do not fully understand it today. However, it was realized in the 1970s that superconductivity comes to the rescue again, giving an understandable explanation of how confinement can happen. An isolated magnetic monopole would have infinite energy in a superconductor because of the Meissner effect. As sketched in Fig. 12, a monopole is a source of magnetic flux, but the Meissner effect would cause this flux to be compressed into an Abrikosov–Gorkov flux tube, with an energy proportional to its length.

The best qualitative understanding that we have of confinement today is to say that it involves a "dual" to the Meissner effect, where this "duality" somehow generalizes to non-abelian gauge theory the symmetry of Maxwell's equations that exchanges electric and magnetic fields. Electric charges — quarks — are then confined by a "dual Meissner effect." We do not fully

Flux Tube

Fig. 12.   In a superconductor, a magnetic monopole (small ball) is the source of a flux tube. As a result, its energy grows linearly with the size of the system.

understand this in the case of QCD, but by now we know various situations in four-dimensional gauge theories in which something like this happens.

Now we come to the weak interactions. The original Weinberg–Salam model was based on a weakly-coupled picture with an elementary Higgs field — an elaboration of the Landau–Ginzburg and Higgs models to include non-abelian gauge symmetry and leptons (and later quarks). But many physicists for decades have wondered if the analogy with superconductivity is even stronger — if the breakdown of the electroweak gauge symmetry involves something more like the BCS mechanism of superconductivity.

There have been numerous motivations, and of course different physicists have had different motivations at different times. Some simply suspected that the analogy between the weak interactions and superconductivity would turn out to be even closer. Some considered the model with an elementary scalar field to be arbitrary and inelegant. Another motivation for some was the fact that the Standard Model is not predictive for lepton and quark masses (and "mixing angles"). Each mass is a free parameter, determined by the strength of the coupling of the Higgs field to a given quark or lepton. Maybe a model of "dynamical symmetry breaking," more like the BCS mechanism, would give a more predictive model.

Perhaps the most compelling motivation came from the "hierarchy problem." Although the electroweak gauge theory with a Higgs field is renormalizable, there is a puzzle about it. In the action describing the Higgs field

$$\int \mathrm{d}^4 x (D_\mu \bar\Phi D^\mu \Phi - \lambda(|\Phi|^2 - a^2)^2),$$

the parameter a^2, which determines the mass scale of weak interactions, is a "relevant" parameter in the renormalization group sense. Generic ideas of renormalization theory suggest that a^2 should be in order of magnitude as large as the largest mass scale of the theory — probably the mass scale of gravity or of grand unification of some sort, but anyway much bigger than the mass scale of weak interactions. By analogy, in condensed matter physics, unless one tunes a parameter — such as the temperature — one does not see a correlation length much longer than the lattice spacing. Why the electroweak length scale is so much bigger than the particle physics analog of the lattice spacing is the "hierarchy problem."

There is no problem writing down a model that replaces the Higgs field with a pairing mechanism (involving a new "technicolor" gauge symmetry with "techniquarks") and solves the hierarchy problem. There is even an immediate success: such a model can easily reproduce a relationship between the W and Z masses and the weak mixing angle that was one of the early triumphs of the Standard Model. However, a serious problem was well-recognized in the late 1970s: one can argue that the way the Standard Model gives quark and lepton masses is inelegant and unpredictive, but at least it works. Simple models of "dynamical electroweak gauge symmetry breaking" have serious problems giving realistic quark and lepton masses. Of course, people found clever fixes but it never looked like a match made in heaven.

Experiment began to weigh in seriously in the 1990s. Neither the Higgs particle nor the new particles required by "dynamical" models were discovered. But tests of the Standard Model — especially in e^+e^- annihilation — became precise enough that it was possible to say that the original version of the Standard Model with a simple Higgs field is a better fit than more sophisticated "dynamical" models. There certainly were still fixes, but people had to work harder to find them.

Probably we all know where this story has reached, at least for now. A particle with properties a lot like the Higgs particle of the Standard Model was found a year ago with a mass around 125 GeV. It looks like the electroweak scale is weakly coupled, as is possible in part because of Anderson's insights about gauge symmetry breaking in 1962. But the hierarchy problem is still with us. "Dynamical" models that tried to solve it have not been confirmed, and weakly coupled models — notably based on supersymmetry — that tried to solve it have also not yet been confirmed. I will just end with a question: When the LHC gets to higher energies in 2015, will this situation persist or will it be resolved?

A Short History of the Theory and Experimental Discovery of Superfluidity in ³He*

W. F. Brinkman

Princeton University

I discuss the development of the theory and experiments on superfluid ³He. After the discovery of superfluidity in ³He by Osheroff, Richardson and Lee, Phil Anderson quickly recruited Doug Osheroff to come to Bell Labs and set up a dilution fridge to continue his experiments. One of the mysteries at that time was how the high-temperature A-phase, which has a gapless excitation spectrum, could be stabilized relative to the fully gapped, lower temperature B-phase. I explain how Phil Anderson and I developed the spin fluctuation theory of the A-phase of superfluid ³He which accounted for its stability, leading to the Anderson–Brinkman–Morel (ABM) theory of the superfluid A-phase.

Very shortly after the publication of the BCS theory of superconductivity a number of authors started speculating on the possibility of superfluidity in ³He. After all it is a Fermi liquid and if the transition were to take place it experimentally had to be well below the Fermi energy and a weak coupling theory should be appropriate. Anderson was among the first to realize this possibility although there were several papers all published in 1960 that made the same suggestion. However, it was the extended paper by Anderson and Morel (1961) that did the first calculations of the possible order parameter and its various properties. At the time d-wave pairing was considered more likely and they concentrated more on that possibility but also worked out a number of results of p-wave pairing in which the up and down spins were simply paired separately.

---

*For a more complete review and discussion of the history of the subject see, P. W. Anderson and W. F. Brinkman in *The Physics of Liquid and Solid Helium*, Part II (John Wiley and Sons, 1978).

A few years later, Balian and Werthamer (1963) gave the first proper treatment of all the spin components and showed that in fact there was a more symmetric solution involving all three components of the $S = 1$ spin triplet, $M_s = \pm 1$ and 0. This state possessed a gap over the entire Fermi surface and a magnetic susceptibility reduced by $1/3$ due to the presence of pairs having $M_s = 0$. In 1965, Leggett showed that the strong interaction parameters of the Landau Fermi Liquid theory would have a large effect on the susceptibility and specific heat. In the case of the susceptibility its value would likely be $1/3$ of the normal state value instead of $2/3$.

A final contribution before the experimental discovery came from the ideas regarding spin fluctuations of Berk and Schrieffer. Layzer and Fay examined the possible effects of spin fluctuations on the transition temperatures for both singlet and triplet pairing. They showed that spin fluctuations strongly suppress singlet BCS type pairing while enhancing triplet p-wave pairing. Thus it appeared from theory that if superfluidity were to be observed in ^{3}He it would be a p-wave state.

This was the status until in 1972 when Osheroff, Richardson and Lee made the first experimental observations of possible superfluid behavior in ^{3}He. Phil was, of course, immediately excited by their results and he recruited Doug Osheroff to come to the Bell labs and quickly set up a dilution refrigerator to do further experiments on the new phases. This led to many collaborations among people at Bell including myself, Doug and Phil but also with other theorists such as M. C. Cross, E. I. Blount, H. Smith and S. Engelsberg. On the experimental side Doug collaborated with L. Corruccini in a number of experiments.

The experimental observations were of various shifts in the nuclear resonant frequencies in the liquid. Leggett quickly developed a theory of how these shifts could arise from superfluid pairing. His work became the basis for essentially all future NMR experiments and led to his receiving the Nobel Prize in 2003. The experimental data suggested was that in fact there were two different phases exhibited by the fluid as a function of temperature. The early experiments, performed along the melting curve, exhibited a high-temperature phase below 2.6 mK that had a BCS-like change in the specific heat and unchanged susceptibility. A second phase appeared as a first order transition at 2.0 mK into a state with a reduced susceptibility. Given the fact that the BW state had an energy gap at all points on the Fermi surface it was surprising that another phase could compete with it.

During those days Phil and I had our offices in the same hall at Bell Labs and we often had coffee in one or the other's office and discussed various aspects of physics. I was aware of the paper by Layzer and Fay that showed how spin fluctuations could enhance the transition temperature of p-wave pairing in Fermi systems and I mentioned this to Phil. He started asking questions about this paper and seemed to get more and more interested. Then one day he came into my office and put a draft of a paper on my desk and said why don't we publish this?

I was not sure how to react to this as I had not really contributed except to tell Phil about the results of Layzer and Fay. I decided that I would try to make sure that I believed the paper so I did a simple Landau/Ginsberg formulation of the problem to see if the result of the spin fluctuation calculation made sense. At the time we were using the Balian–Wertheimer formulation of the problem and an Anderson–Morel state that they said was the same as the one originally used. It turned out that within mean field theory this version of Anderson–Morel state was always unstable and therefore could not possibly be the explanation of the observed higher temperature phase. By this time Phil had left for Cambridge, splitting his time between the Cavendish Lab and Bell Labs. So the only sensible thing to do was to call Phil and give him the bad news. When I got him on the phone and told him what I had learned he quickly said "Now I understand the difference between the state used by Balian and Wertheimer and the actual Anderson and Morel state we used. For a minute I did not understand then I realized he was telling me to use the original state that Anderson and Morel had used and that would solve the problem. How he could know this is still a mystery to me to this day but, in fact, when I went back and looked it was true that the original state was stable and was the natural candidate for the A phase of superfluid ^{3}He. The rest is history. The state became known as the ABM state and much has been learned about its properties.

To me this kind of insightfulness is the essence of what has been great about Phil Anderson and I am sure many others have similar stories.

A lot of work went into understanding the spin fluctuation theory better and to incorporate it into the Landau Fermi liquid theory. The latter was done by Serene and Rainer who came up with an expansion in powers of the ratio of the transition temperature to the Fermi temperature.

In order to explain the two states further I need to introduce the nature of the order parameter of a state with p-wave pairing. The order parameter in a p-wave superfluid is a complex matrix usually defined as

$$\Delta = \sum_{i=1}^{3} \sigma_i \sigma_y d_i(\mathbf{k}) = \sum_{ij=1}^{3} \sigma_i \sigma_y d_{ji} k_j.$$

Here the σ_i are the Pauli matrices and $\mathbf{k}$ is the unit vector in reciprocal space. The matrix d_{ji} is a 3×3 matrix which is the basic order parameter. Since we are discussing p-wave pairing the order parameter depends linearly on the unit vectors in $\mathbf{k}$ space and must be a triplet in spin space. It also must have the feature that in the absence of spin-orbit coupling the frame in which the orbital components are defined is independent of the frame that defines the spin components.

The state proposed by Balian and Werthamer is a completely symmetric state

$$d_{ji} = d \begin{pmatrix} 1 & 0 & 0 \\ 0 & 1 & 0 \\ 0 & 0 & 1 \end{pmatrix},$$

while the Anderson (Brinkman) Morel state is

$$d_{ji} = d \begin{pmatrix} 1 & 0 & 0 \\ i & 0 & 0 \\ 0 & 0 & 0 \end{pmatrix}.$$

This state is different than the BW version which has

$$d_{ji} = \begin{pmatrix} 1 & 0 & 0 \\ 0 & 1 & 0 \\ 0 & 0 & 0 \end{pmatrix},$$

which, as mentioned earlier, can be shown to always be unstable to small perturbations. The fundamental reason that this state is unstable has to do with the fact that the up and down spin states have the orbital angular momentum in opposite directions, a very unlikely situation.

The result of the spin fluctuation calculation was that the ABM state was stable if the spin fluctuations were sufficiently strong. This was consistent with the experimental data. Since then an enormous number of NMR, specific heat and susceptibility measurements have all verified this prediction.

In order to think about these experiments it was essential to simplify the order parameter so that one can visualize what is happening. For the ABM state the order parameter is defined by the direction of the orbital angular momentum of the pairs usually labeled $\mathbf{L}$ and the direction along which the spins are in an $M_S = 0$ state usually designated by $\mathbf{k}$. The BW state is specified by the difference between the reference frame of the spin states relative to that of the orbital degrees of freedom. Thus the order parameter is specified by the axis of rotation and the magnitude of the rotation between the two reference frames usually written as **n** and angle θ. Spin-orbit coupling, although very weak, forces the **L** and **k** vectors to be parallel and a magnetic field forces them to be perpendicular to it. This is the equilibrium configuration in the bulk of the ABM liquid. Likewise for the BW state the spin-orbit coupling forces the angle of rotation θ to be 104° while the magnetic field wants the rotation axis to be parallel to it. These configurations along with boundary conditions at surfaces allowed one to predict and calculate the results of various experiments using NMR and Phil contributed many ideas and suggestions to this work. They allow one to understand possible textures of the order parameters and the itemization of the possible defects in each phase.

Today these results are well-established and more exotic effects such as bound states below the continuum of particle-hole excited states are being observed. There has been much speculation that there might be p-wave metals but the only one identified so far is the ruthenate, Sr_2RuO_4. Of course, the high-temperature superconductors are known to be d-wave states but because they are two-dimensional in a highly anisotropic lattice they do not exhibit the fascinating textures and NMR effects that are present in ^{3}He. Recently a p-wave state is believed to occur in a linear chain of Iron atoms where the p-wave superconductivity is induced by the underlying superconducting metal. These states are much harder to do detailed experiments on than ^{3}He. Thus ^{3}He has been a unique laboratory for exploring a p-wave angular momentum state. There were many contributors to understanding of these states but Phil was certainly one if not the most important contributor.

Superconductivity in a Terrestrial Liquid: What Would It Be Like?

A. J. Leggett

Department of Physics
University of Illinois at Urbana-Champaign
Urbana, IL 61801, USA
aleggett@illinois.edu

> Although it seems rather unlikely that superconductivity could occur in the liquid state under ambient conditions, there seems to be no rigorous principle which forbids it. I raise, and make a first pass at answering the question: What would be the anomalous macroscopic properties of such an "ambient-conditions liquid superconductor"?

It is a pleasure to have been invited to contribute to this volume which celebrates the 90th birthday of the doyen of contemporary condensed matter theory, Phil Anderson. I thought I would take the opportunity to indulge in a little piece of science-fiction fantasy, analogous to what was done by the late novelist Kurt Vonnegut in his invention of ice-IX.[1] Readers may recall that at the time that his novel *Cat's Cradle* was written, there were eight different known solid phases of water, none of them stable above zero temperature Celsius. Vonnegut (whose brother was a physical chemist) postulated a ninth form, which would be stable at and above room temperature, but which was separated from both liquid water and the known phases of ice by a free energy barrier so high that it had never been realized in the history of the earth. One day a scientist synthesizes it in his laboratory, and it is eventually released into the environment. What Vonnegut implicitly assumes, but does not tell the reader (he is writing a novel, not a physics textbook!) is that the transition between water and ice-IX belongs to the small sub-class of first-order transitions (which includes the A-B transition in superfluid liquid helium-3)[2] which are "hypercooled," that is, which have the property that the latent heat released is insufficient to warm the system back above the equilibrium transition temperature, so that the velocity of propagation is not, as in the usual case, limited by the need to get rid of this heat but only by the speed of sound. As a result, when the ice-IX sample is released, the oceans freeze near-instantaneously, with predictably gloomy consequences for mankind. Of course, as far as we know, ice-IX is (thankfully!) a fiction, but by thinking about it one learns quite a bit more generally about the properties of first-order phase transitions.

The fantasy I am going to explore is equally improbable but also perhaps equally fertile in its implications: a liquid which is superconducting under conditions which are not too remote from the ambient terrestrial ones (e.g., similar to those under which existing high-temperature superconductivity occurs in the cuprates). The present context may not be totally inappropriate for such an exploration, since it makes contact with at least three of the many areas in which Phil has worked, namely high-temperature superconductivity, neutron stars (see below) and the anomalous electrostatic effects discovered by Tao and co-workers in ordinary superconductors.[3] The main question I will raise is: what would be the novel macroscopic effects associated with such a system? An attempt to answer this question may perhaps give us a new perspective on the "familiar" superconductivity which occurs in solids.

Before addressing this question, however, a few comments are in order. First, the idea of superconductivity occurring in a system which by any reasonable definition is liquid is of course itself not at all novel: it is confidently believed that in certain regions of a neutron star not only the neutrons but the (minority) protons may form Cooper pairs, and thus the system would be expected to show not just superfluidity but also superconductivity, with the usual consequences such as magnetic vortices. Closer to home, Ashcroft and co-workers have conjectured[4] that hydrogen, when subjected to sufficiently high pressures ($\sim$a few $100\,\mathrm{GPa}$) might not only become a metallic liquid but also form Cooper pairs, possibly of both electrons and protons, and thus exhibit superconductivity; they have moreover raised the question how one might identify such an anomalous superconducting state.[5] However, the attainment of such high pressures would presumably require experimental conditions very different from those under which the conventional superconductors are usually studied, so that these authors do not place much emphasis on any novel features in the gross macroscopic behavior, but concentrate mainly on effects associated with the coexistence of two different order parameters. Thus, as far as I know, the question I raise here has not previously attracted much attention in the literature.

Secondly, let's just think for a minute about how improbable a liquid superconductor under "near-ambient" conditions (hereafter abbreviated ACLS to stand for Ambient Conditions Liquid Superconductor) actually is.[a] To the best of my knowledge, while there may be electrolyte solutions which are liquid down to the highest temperature at which superconductivity is currently known to occur (about 160K), the lowest temperature at which any liquid metal phase is stable under atmospheric pressure is 234K (for Hg), leaving some distance to go. Of course, this simple comparison may not be very meaningful. Indeed, we should consider separately the prospects for "BCS-like" (phonon-mediated) superconductivity and for the "exotic" (all-electronic) type. As regards the former, while the atomic disorder characterizing the liquid state may not in itself be an impediment to the occurrence of superconductivity (some of the highest-temperature BCS superconductors are strongly amorphous alloys), it is not entirely clear whether the fact that the disorder is time-dependent, with a timescale which is of the same order as or shorter than the typical timescale for superconductivity (Planck's constant divided by the thermal energy at the transition temperature) would have a strongly inhibiting effect on the latter (note that this feature of liquid behavior is not entirely captured by the structure of the phonon spectrum as studied in Jaffe and Ashcroft[4]). My untutored guess is that it would not; however, since among the known superconductors the ones with the highest transition temperatures are "all-electronic", one might *prima facie* think that an ACLS would be likely to use the latter mechanism. Here, however, there is a serious difficulty: the currently available experimental evidence suggests that almost without exception the order parameter in these existing materials is anisotropic (non-s-wave). As is well known, such an order parameter is highly vulnerable to even static disorder, let alone the time-dependent variety. Thus, all in all, the existence of an ACLS seems rather improbable (but then, so prior to 1986 did superconductivity above 100K!).

I now turn to the main topic of this essay, namely the novel macroscopic properties that the ACLS state might be expected to manifest. While the discussion below is entirely qualitative, for definiteness I shall assume that the (orders of magnitude of) the physical parameters of the ion and electron systems are comparable to those of (say) the cuprates, so that, for example, the superconducting condensation energy is of the order of a few degrees

---

[a] After submission of the manuscript, I became aware of the paper of P. P. Edwards *et al.*, *ChemPhysChem* **7**, 2015 (2006), which discusses the possible existence of ACLS in liquid metal-ammonia solutions.

K per (superconducting) electron. For comparison, the energy associated
with surface tension should be of order 10^4K per (surface) atom (c.f. below),
while that required to lift an atom through 1 cm in the earth's gravitational
field is about 1 mK. Armed with these numbers, let us consider the likely
behavior of the system under two different kinds of experimental conditions:
(a) the ACLS (or more precisely the normal liquid metal just above the
notional onset temperature for superconductivity) is confined to a cell or
tube, with no free surface, (b) it is contained in an open bowl or similar
geometry. In contrast to Jaffe and Ashcroft,[4] I will assume throughout that
pairing occurs only in the electron system, the ions themselves remaining
throughout perfectly "normal".

Let's first consider case (a). In the case of the traditional (solid)
systems, superconductivity manifests itself at the macroscopic level prin-
cipally through two phenomena, the Meissner effect and the near-infinite
metastability of circulating currents in a ring geometry (the more easily
demonstrated phenomenon of nonzero current through a wire with zero
voltage drop can be reduced, conceptually speaking, to one or the other of
these two depending on the parameters). The Meissner effect is a thermodynamic equilibrium phenomenon, and so *prima facie* the only way a difference
between a standard superconductor and an ACLS could show up would be if
the ion system could deform appreciably to give a lower-energy equilibrium
state; since compressional energies are presumably similar to those in the
standard case, this seems unlikely in the closed geometry of case (a). So I
think my *prima facie* expectation would be that the ACLS would show a
complete Meissner effect in a sufficiently small external magnetic field, and
also the conventional type-II behavior in higher fields (since there seems no
obvious reason why the standard Abrikosov vortices should not occur, with
static properties similar to the usual case).

With regard to the possibility of persistent supercurrents, which is a
metastable rather than a thermodynamic equilibrium phenomenon, the most
obvious difference with a very disordered solid alloy is that, at least over
timescales long compared to the typical ionic rearrangement time, there is no
obvious way in which any Abrikosov vortices present are going to be pinned;
hence one would expect that as soon as the field generated by the current
exceeds the lower critical field (which is likely to be of the order of the earth's
magnetic field) the system will transition to a resistive state. This behavior
is not of course qualitatively different from that of some known superconductors. However, there is a second possible mechanism of resistance which has
no analog in that case: Consider a thin-ring geometry (transverse dimensions

of the ring smaller than the pair radius), and the nature of a fluctuation which will allow the order parameter to slip 2π of phase and thereby reduce the circulating supercurrent. In the usual Langer–Ambegaokar–McCumber–Halperin ("LAMH") scenario[6,7] the mechanism is a fluctuation to zero of the order parameter over a length of the order of the pair radius, with the *total* electron density remaining close to constant over the whole length of the phase slip; such a process has a cost per unit area of the order of the condensation energy times the pair radius. Any large deviation in the total electron density, such as being driven to zero over this or a smaller length, is strongly suppressed by the Coulomb force. However, in an ACLS the ions may be able to move so as to counteract this effect, and the only extra energy involved is then that involved in the surface tension necessary to form a thin slab of vacuum bridging the cross-section of the ring, thereby allowing the phase slip. Given the order-of-magnitude numbers listed above, it seems possible that this mechanism might be competitive with the LAMH one, thus possibly rendering unstable superflows in thin rings which in the standard scenario would be effectively metastable. In thicker rings the absence of a vortex pinning mechanism is likely to have a qualitatively similar effect, cf. above.

Apart from the two fundamental manifestations of superconductivity discussed above, the traditional systems show a variety of other characteristics: vanishing Peltier coefficient, Josephson effect, etc. To the extent that the leads themselves are "conventional" normal or superconducting materials (which the liquid ions cannot penetrate) it seems likely that the qualitative behavior of an ACLS with respect to such phenomena will be similar to the normal one.

Things become a lot more interesting when we consider the "open" geometry of case (b). Those readers old enough (like the present author) to remember playing in their pre-OSHA (Occupational Safety and Health Administration) childhoods with droplets of mercury in the kitchen sink will not need to be reminded that the surface tension of metallic liquids is exceptionally high, ~ 1 in SI units (which translates into the figure given above). However, it is conceivable that this is not the only germane consideration: in particular the "Tao effect"[3] may be relevant. This effect is most naturally interpreted as showing that at least under certain circumstances there is an extra surface energy associated with the superconducting state. Originally, the effect was thought to be peculiar to the cuprates, and an explanation was developed[3] in which a crucial role was played by the strong anisotropy of these materials, a feature which would presumably be lacking in an ACLS; however, subsequent experiments[8] suggest that it does not require such anisotropy but is actually a generic property of the superconducting state. To be sure, in existing experiments the surface energy in question is small ($\sim 10^{-3}$ SI), and occurs only in electric fields $\sim 1\,\mathrm{kV/cm}$, but the mere fact that it is not fully understood indicates a non-negligible possibility that the surface tension of an ACLS may be anomalous even in the context of liquid metals.

Irrespective of this (but assuming the surface tension is not much *less* than that of a typical liquid metal), let us consider how an ACLS may be expected to behave in a weak magnetic field (such as that of the earth). A crucial difference from the case of an ordinary solid superconductor is of course that the electron and ion systems can deform together, thus preserving overall charge neutrality and avoiding the activation of the strong Coulomb forces. Thus, we need to minimize the sum of at least four and possibly five different energies: condensation, magnetic, ordinary surface tension, gravitational and possibly anomalous electrostatic ("Tao") energies. This would seem to be a rather non-trivial (and highly geometry-dependent) problem. Were it not for the large surface tension contribution, my gut instinct (not based on any quantitative calculation at this stage) is that the ACLS would spontaneously form a thin film and coat its surroundings (thereby presumably constituting a considerable safety hazard!). Presumably, the surface tension would suppress this behavior in its extreme form, but for a large enough sample this effect must be overwhelmed by the terms proportional to volume. So, if we imagine trying to conduct a standard levitation experiment with the ACLS originally contained in an open bowl, what will happen? Readers are invited to make their own guesses/calculations; all I know is that I would not like to be the environmental safety officer responsible for developing a handling protocol for this system.

Obviously, while I have stated in this essay what I believe to be an interesting problem, the above discussion only scratches its surface, and is almost certainly lacking in sufficient imagination. Indeed, I would take a large bet that in the improbable event that an ACLS is actually realized in the laboratory, it will rapidly turn out to have various novel and intriguing properties not anticipated above. I heard the story (for whose authenticity I cannot vouch) that when the Finnish electronic giant Nokia had produced a new type of hand-held electronic device and wanted to explore its potential, it let loose on it a group of seven-to-nine-year-olds, who rapidly came up with applications which their elders had never imagined. Perhaps we should do the same with an ACLS if it is ever realized!

Acknowledgments

This material is based upon work supported as part of the Center for Emergent Superconductivity, an Energy Frontier Research Center funded by the United States Department of Energy, Office of Science, Office of Basic Energy Sciences under Award number DE-AC0298CH1088. I thank Cai Peng and Tang Peizhe for helpful comments. It is a pleasure to dedicate this essay to Phil Anderson and to wish him many more years of happy and fruitful research in physics.

References

1. J.K. Vonnegut, *Cat's Cradle* (Delacorte Press, New York, 1963).
2. A.J. Leggett and S.K. Yip, in L.P. Pitaevskii and W.P. Halperin (Eds.), *Helium-3* (North-Holland, Amsterdam, 1990).
3. R. Tao, X. Zhang, X. Tang and P.W. Anderson, Formation of high-temperature super-conducting balls, *Phys. Rev. Lett.* **83**, 5575–5578 (1999).
4. J.E. Jaffe and N.W. Ashcroft, Superconductivity in liquid metallic hydrogen, *Phys. Rev. B* **23**, 6176–6179 (1981).
5. E. Babaev, A. Sudbø and N.W. Ashcroft, Observability of a projected new state of matter: a metallic superfluid, *Phys. Rev. Lett.* **95**, 105301 (2005).
6. J.S. Langer and V. Ambegaokar, Intrinsic resistive transition in narrow superconducting channels, *Phys. Rev.* **164**, 498–510 (1967).
7. D.E. McCumber and B.I. Halperin, Time scale of intrinsic resistive fluctuations in thin superconducting wires, *Phys. Rev. B* **1**, 1054–1070 (1970).
8. R. Tao, X. Xu, Y.C. Lan and Y. Shiroyanagi, Electric-field induced formation of low temperature superconducting granular balls, *Physica C* **377**, 357–361 (2002).

40 Years of Quantum Spin Liquid:
A Tale of Emergence from Frustration

Patrick A. Lee

Department of Physics, MIT
Cambridge, MA 02139, USA
palee@mit.edu

On the occasion of the celebration of Phil Anderson's 90th birthday, and the 40th anniversary of his introduction of the spin liquid concept, I give an overview of the current status of quantum spin liquids. With the discovery of several experimental examples, quantum spin liquid has become the poster child of the notion of emergence, i.e. novel particles and gauge fields emerge in the low energy effective theory and dominate the low energy physics. In addition to geometrical frustration, the route to spin liquid includes the proximity to the Mott transition. The seed planted by Phil 40 years ago has indeed blossomed and prospered.

Louis Néel proposed the possibility of antiferromagnetic order back in 1936,[1] and by now Néel order is ubiquitous and seems second nature to condensed matter physicists. The possibility of quantum fluctuations destroying Néel order was raised in 1973 by P. W. Anderson.[2] He focused on the singlet formation between two spins with energy gain of $-S(S+1)J$. The origin of the factor unity in $(S+1)$ is quantum mechanics and $S = \frac{1}{2}$ enjoys the greatest energy gain. In a spin chain, a trial wave function of singlet dimers yields an energy of $-\frac{3}{8}J$ per bond, which is lower than the Néel value of $-\frac{1}{4}J$. Indeed, we now understand Bethe's solution as a linear superposition of singlets with varying range. Anderson suggested that a similar superposition may hold for higher dimensions, which he called the resonating valence bond (RVB) state. He further proposed that frustrated lattices such as the triangular lattice may help stabilize the RVB state vs. Néel order. Unfortunately it was found that the Heisenberg model can partially relieve frustration by forming 120 degrees order.[3] The field laid dormant until 1987, when soon after the discovery of high-temperature superconductors, Anderson argued that upon doping, the RVB state naturally forms a superconductor because the singlet pairs are now mobile and can be viewed as Cooper pairs.[4] This suggestion launched a great deal of theoretical activities. In particular, there has been a great deal of development in the "easier" problem of quantum disordered spin states, which we call spin liquids. The requirement for a spin liquid is that it is a charge insulator with an odd number of electrons per unit cell (i.e. a Mott insulator) which does not exhibit Néel order down to zero temperature, despite the presence of antiferromagnetic exchange. We now know that the spin liquid is more than just the absence of order. Theory predicts the emergence of new excitations such as particle carrying $S = \frac{1}{2}$ and no charge, called spinons. Spinons may be fermions or bosons, and the excitation may gapped or gapless. Furthermore, the spinons do not live by themselves, but are coupled to an emergent gauge field, which may be $U(1)$ or $Z2$. Current analytic theory is not powerful enough to tell us which of these possibilities is realized for a given Hamiltonian, but the different possibilities can be classified and their nature fully described. As we shall see, the spin liquid problem is closely related to lattice gauge theory coupled to matter field, and the spin liquid states correspond to the deconfined phases.

The field of spin liquids received a powerful boost several years ago with the discovery of promising experimental candidates. They include the organic molecular compounds and a mineral called Herbertsmithite, which is a $S = \frac{1}{2}$ Kagome lattice that can be synthesized in the laboratory.[5] The $S = \frac{1}{2}$ Kagome lattice is a highly frustrated two-dimensional lattice which has long been suspected to harbor a spin liquid ground state based on numerical work.[6] What is interesting is that in addition to using frustration to stabilize the spin liquid as proposed by Anderson, the organic materials have taught us a second route, i.e., the proximity to the Mott transition. The two families of organic material consist of molecular dimers (called ET and dmit as abbreviation) which carry a single electron each, and form an approximate triangular lattice. It can be modeled as a Hubbard model on a triangular lattice with one electron per unit cell.[7,8] Deep in the Mott insulator regime, the low energy physics is described by a nearest-neighbor Heisenberg model and the system is expected to order into the 120 degree phase. The spin liquid materials turn out to be insulators which turn into a metal (dmit) or superconductor (ET) under modest pressure. The proximity to the Mott transition means that while the low energy excitations may still be described by the spin degrees of freedom, the Hamiltonian is much more complicated than the nearest-neighbor Heisenberg. For example, ring exchange terms may be important.[7] We therefore focus on the nature of the ground state, and leave aside the question of what spin Hamiltonian can stabilize it.

These two quite distinct spin liquid candidates share the common feature that they both appear to be gapless. In the organic system, the spin susceptibility goes to a constant and the specific heat has a linear T term,[9] both characteristic of a metal with a Fermi surface, and most uncommon for an insulator. Thermo-conductivity κ was measured and it is found at least for the dmit salts that κ/T goes to a constant at low temperature.[10] All these strange properties are consistent with the emergence of fermionic spinons which form a Fermi surface. The situation with Herbertsmithite is less clear because there exists a significant amount of local moments between the Kagome planes which obscure the low temperature behavior. Recently a single crystal large enough for neutron scattering was grown.[11] The data show that the spin excitation is gapless and rather featureless as a function of energy and has been argued to support the picture of gapless spinons. There are a number of additional examples of spin systems which do not order down to low temperatures. A partial list includes the hyperkagome which is a three-dimensional network of corner sharing triangles,[12] some pyrochlore (corner sharing tetrahedra) structures which are describes as quantum spin ice,[13] and frustrated Mo clusters in $LiZn_2Mo_3O_8$.[14] Further experiments on these and other examples will surely yield new information and surprises.

Now we describe how the Heisenberg model $H = J\sum_{\langle ij\rangle} \boldsymbol{S}_i\cdot\boldsymbol{S}_j$ is mapped onto a lattice gauge theory. It is convenient to introduce fermions $f_{i\alpha}$ which carry spin index α on site and write the spin operator as

$$\boldsymbol{S}_i = \frac{1}{2} f_{i\alpha}^\dagger \sigma_{\alpha\beta} f_{i\beta}. \tag{1}$$

The fermions must be subject to the constraint

$$f_{i\uparrow}^\dagger f_{i\uparrow} + f_{i\downarrow}^\dagger f_{i\downarrow} = 1, \tag{2}$$

which excludes empty or doubly occupied fermions. A similar representation can be made by introducing spin carrying bosons called Schwinger bosons instead of fermions. However, with bosonic representations, the mean field theory yields either Néel order or gapped spin liquid states and gapless spin liquid requires fine tuning to a quantum critical point. On the other hand, we will see that fermionic representation naturally accommodates gapless phases. Since the experimental systems observed so far appear to be gapless, we will use the fermionic representation. The Heisenberg term is quartic in the fermion operator and is given as[15]

$$S_i \cdot S_j = -\frac{1}{4} f_{i\alpha}^\dagger f_{j\alpha} f_{j\beta}^\dagger f_{i\beta} - \frac{1}{4} \left(f_{i\uparrow}^\dagger f_{j\downarrow}^\dagger - f_{i\downarrow}^\dagger f_{j\uparrow}^\dagger \right) \left(f_{j\downarrow} f_{i\uparrow} - f_{j\uparrow} f_{i\downarrow} \right) + \frac{1}{4} f_{i\alpha}^\dagger f_{i\alpha}$$

$$(3)$$

Eq. (2) invites mean field decoupling in the particle-hole and particle-particle channels

$$\chi_{ij} = \langle f_{i\alpha}^\dagger f_{j\alpha} \rangle \quad \text{and} \tag{4}$$

$$\Delta_{ij} = \langle (f_{j\downarrow} f_{i\uparrow} - f_{j\uparrow} f_{i\downarrow}) \rangle. \tag{5}$$

For simplicity, let us ignore the pairing channel and consider χ_{ij} only. Note that at the mean field level, the fermion acquires dynamics and is now free to hop on the lattice. Formally we introduce the Hubbard–Stratonovich field χ_{ij} and the Lagrange multiplier λ_i to enforce the constraint Eq. (2).

$$Z = \int d\chi d\lambda df df^\dagger e^{-S} \tag{6}$$

$$S = \int d\tau \left[\sum_i \left[f_{i\alpha}^\dagger \partial_\tau f_{i\alpha} + i\lambda_i \left(f_{i\alpha}^\dagger f_{i\alpha} - 1 \right) \right] \right.$$

$$\left. + \sum_{\langle ij \rangle} J \left[\chi_{ij} \left(f_{j\alpha}^\dagger f_{i\alpha} + \text{h.c.} \right) + 2|\chi_{ij}|^2 \right] \right]. \tag{7}$$

The saddle point describes the mean field theory. Fluctuations around the saddle point are dominated by the phase fluctuation of χ_{ij} which becomes the compact $U(1)$ gauge field a_{ij} on link ij while $i\lambda$ becomes the time component. The low energy theory of Eq. (1) becomes a compact $U(1)$ lattice gauge theory coupled to fermions hopping on a lattice.[16]

It is not surprising that a gauge theory should result from a representation subject to constraints. Because the fermions introduced extra degrees of freedom, we see that there is a gauge redundancy in Eq. (1), i.e., S is invariant under a local gauge transformation $f_i \rightarrow f_i e^{i\phi}$. Therefore a gauge theory must emerge.

Several mean field solutions are notable. If χ_{ij} is a constant, we will have a half-filled band with a Fermi surface. Another interesting class is called the flux phases where nonzero gauge fluxes penetrate the plaquettes. π-flux for the square lattice[17] and π flux through the hexagons in the Kagome lattice yield massless Dirac fermions.[18] Note the time reversal symmetry is preserved with π fluxes. If we include pairing we can gap the Fermi surface in a variety of ways. We do not know which mean field theory is closer to the truth for a given Hamiltonian, but we can discuss the consequences of each state and compare with experiments. This is why the availability of experiments is crucial for the advancement in the field. Recent advances in numerical methods such as DMRG[19] and projected wave functions plus Lanczos[20] are making important contributions to ascertain the domain of stability of spin liquids for a given Hamiltonian.

The enemy of the spin liquid is confinement. If the gauge theory is in a confined phase, the low lying degrees of freedom remain those of spin excitations, and the system typically breaks translational symmetry, either by Néel order or by forming a valence bond solid. On the other hand, if we are in the deconfined phase, the fermion and gauge fields emerge as new particles and fields at low energies. The fictitious fermions which we introduced as a formal device and had no dynamics at high energy and short distance take on a life of their own at low energies. This is a prime example of the notion of emergence. There are now several exactly solvable examples which show that this scenario plays out in certain spin Hamiltonians.[21,22] The reader should not be concerned that the fermions are not gauge invariant. After all the electrons in our world are also not gauge invariant but we have no trouble thinking of them as real objects. The difference between electron and spinon is a qualitative one: the electrons are weakly coupled to the gauge field with a small fine structure constant whereas spinons couple with strength unity. Another difference is that in solids the electron velocity v_F is much less than the speed of light and the coupling to electric field is stronger than the coupling to magnetic field. In spin liquids both velocities are given by J and transverse gauge fluctuations which are unscreened become dominant. For example, for $U(1)$ spin liquids this gives rise to the prediction that the specific heat $\sim T^{2/3}$, i.e., non-Fermi liquid behavior,[23] a subject under intense studies today.[24–27] The possibility of exchanging transverse gauge fluctuations (the analogy of magnetic field) which leads to attraction between spinon moving in the same direction by the Ampere effect has inspired the proposal of Amperean pairing between spinons as a possible explanation of the 6K transition observed in the ET organic compounds.[28]

Pure gauge theory is always confining in two dimensions due to instanton fluctuations.[29] The coupling to matter fields can change that. It was shown that coupling to N components of Dirac spinons can suppress instantons if N exceeds a critical N_c.[30] In the π-flux state $N = 4$, and deconfinement is quite possible. In the case of the Fermi sea, it has been shown that there are so many gapless degrees of freedom that it is analogous to infinite N and deconfinement is always possible.[31]

In conclusion, the field of quantum spin liquid is entering an exciting new phase where dialogs between theory and experiment are possible. We can look forward to exciting discoveries ahead.

I acknowledge the support by the NSF under Grant DMR-1522575.

References

1. L. Néel, *Ann. Phys.* (Paris) **5**, 232 (1936).
2. P.W. Anderson, *Mater. Res. Bull.* **8**, 153 (1973).
3. D.A. Huse and V. Elser, *Phys. Rev. Lett.* **60**, 2531 (1988).
4. P.W. Anderson, *Science* **235**, 1196 (1987).
5. For a review of some of the experimental systems, see P.A. Lee, *Science* **321**, 1306 (2008); L. Balents, *Nature* **464**, 199 (2010).
6. P. Lecheminant, B. Bernu, C. Lhuillier, L. Pierre and P. Sindzingre, *Phys. Rev. B* **56**, 2521 (1997).
7. O. Motrunich, *Phys. Rev. B* **72**, 045105 (2005).
8. S.S. Lee and P.A. Lee, *Phys. Rev. Lett.* **95**, 036403 (2005).
9. S. Yamashita, Y. Nakazawa, M. Oguni, Y. Oshima, H. Nojiri, Y. Shimuzu, K. Miyagawa and K. Kanoda, *Nat. Phys.* **4**, 459 (2008).
10. M. Yamashita, N. Nakata, Y. Senshu, M. Nagata, H. Yamamoto, R. Kato, T. Shibauchi and Y. Matsuda, *Science* **328**, 1246 (2010).
11. T. Han, J. Helton, S. Chu, D. Nocera, J.A. Rodriguez-Rivera, C. Broholm and Y.S. Lee, *Nature* **492**, 406 (2012).
12. Y. Okamoto, M. Nohara, H. Aruga-Katori and H. Takagi, *Phys. Rev. Lett.* **99**, 137207 (2007).
13. K. Ross, L. Savary, B. Gaulin and L. Balents, *Phys. Rev. X* **1**, 021002 (2011).
14. J.P. Sheckelton, J.R. Neilson, D.G. Soltan and T.M. McQueen, *Nat. Mater.* **11**, 493 (2012).
15. G. Baskaran, Z. Zou and P.W. Anderson, *Solid State Commun.* **63**, 973 (1987).
16. G. Baskaran and P.W. Anderson, *Phys. Rev. B* **37**, 580 (1988).
17. I. Affleck and J.B. Marston, *Phys. Rev. B* **37**, 3774 (1988).
18. Y. Ran, M. Hermele, P.A. Lee and X.G. Wen, *Phys. Rev. Lett.* **98**, 117205 (2007).
19. S. Yan, D. Huse and S.R. White, *Science* **332**, 1173 (2011).
20. Y. Iqbal, F. Becca, S. Sorella and D. Poilblans, *Phys. Rev. B* **88**, 060405 (2006).
21. A. Kitaev, *Ann. Phys.* **321**, 2 (2006).
22. X.G. Wen, *Phys. Rev. Lett.* **90**, 016803 (2003).
23. P.A. Lee and N. Nagaosa, *Phys. Rev. B* **46**, 5621 (1992).
24. S.S. Lee, *Phys. Rev. B* **80**, 165102 (2009).
25. M. Metlitski and S. Sachdev, *Phys. Rev. B* **82**, 075127 (2010).
26. D. Mross, J. McGreevy, H. Liu and T. Senthil, *Phys. Rev. B* **82**, 045121 (2010).
27. D. Dalidovich and S.S. Lee, *Phys. Rev. B* **88**, 245106 (2013).
28. S.S. Lee, P.A. Lee and T. Senthil, *Phys. Rev. Lett.* **98**, 067006 (2007).

29. A. Polyabov, *Phys. Lett. B* **59**, 82 (1975).
30. M. Hermele, T. Senthil, M.P.A. Fisher, P.A. Lee, N. Nagaosa and X.G. Wen *et al.*, *Phys. Rev. B* **70**, 214437 (2004).
31. S.S. Lee, *Phys. Rev. B* **78**, 085129 (2008).

High-T_c Superconductivity and RVB

Mohit Randeria

Department of Physics, The Ohio State University
Columbus, OH 43210, USA
randeria@mps.ohio-state.edu

I describe the resonating valence bond (RVB) approach to doped Mott insulators, pioneered by P. W. Anderson, and the insights that it has given into the problem of high-temperature superconductivity in the cuprates. I briefly review our current understanding of the field highlighting both successes and open problems.

1. Introduction

This article is a slightly modified version of the talk I was privileged to give at the Conference at Princeton in December 2013 to celebrate the career of Phil Anderson on the occasion of his 90th birthday. It is perhaps fitting to begin with "More is Different,"[1] one of the most influential essays on the philosophy of science in the past 50 years, which ends by recounting a conversation in Paris in the 1920s. Fitzgerald: "The rich are different from us." Hemingway: "Yes, they have more money." It is equally true that "Phil is different from us" because "he has many more accomplishments."

Here I want to focus primarily on just one of those accomplishments, Phil's 1987 paper[2] on the resonating valence bond (RVB) theory of high-T_c superconductivity (HTSC). According to Google Scholar, this is his second most highly cited paper, second only to the 1958 paper on Anderson localization.

Bednorz and Müller published[3] their discovery of HTSC in $La_{2-x}Ba_x \cdot CuO_4$ in the summer of 1986. By December, several groups around the world had reproduced their experimental results and the scientific community began to take notice of this development. Phil's RVB paper was written in India where he was attending a heavy fermion/mixed valence conference in

M. Randeria

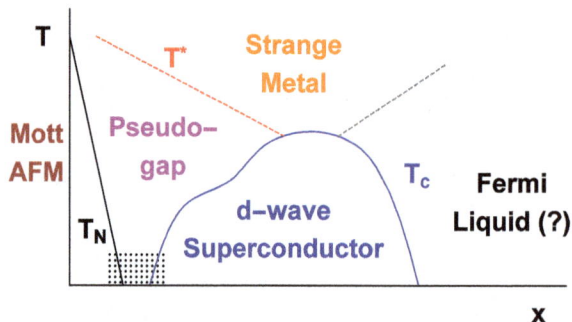

Fig. 1. Schematic phase diagram of the hole-doped high-temperature superconductors in the doping-temperature plane.

early 1987. It was submitted to *Science* in late January and accepted for publication on February 3, 1987, three days *before* the discovery of $YBa_2Cu_3O_7$ was even submitted[4] to *Physical Review Letters*! It is astonishing that at such an early stage in the game, when so little was known about the materials and their properties, the 1987 RVB paper[2]

- proposed an effective low-energy Hamiltonian for the cuprates,
- revived the idea of quantum spin liquid ground state for Mott insulators,
- proposed a radical new vision for superconductivity in doped Mott insulators,
- and speculated on the anomalous properties of the cuprates.

In the remainder of this article, I will discuss each of these points and describe how these ideas have developed over the years. I will also briefly review our current understanding of high-T_c superconductivity in the cuprates and emphasize open questions.

More than a quarter century after the discovery of HTSC, we are still debating many aspects of the physics of cuprates because the problem challenges the central paradigms of 20th century condensed matter physics. First, the undoped, parent insulator represents a failure of band theory with Coulomb correlations giving rise to the Mott insulator. Second, the normal, i.e., non-superconducting, phase is highly abnormal. Landau's Fermi liquid theory, the underpinning of our understanding of conventional metals, fails in both the strange metal and pseudogap regimes of the phase diagram. The electron is not a well-defined excitation above T_c. Third, even the superconducting phase, now known to have d-wave pairing symmetry and supporting well-defined electronic excitations, cannot be described by BCS theory. Its very short coherence length, a superfluid phase stiffness much smaller than the energy gap at low doping, and its insensitivity to disorder, all make the superconducting state different from standard BCS. To add to all this, there are a plethora of order parameters — with broken translational, rotational, or time-reversal symmetry — that compete with superconductivity in the lightly doped Mott insulator and lead to possible quantum critical points.

2. The Hubbard Model

In the 1987 RVB paper, Phil recognized that the layered perovskite structure of La_2CuO_4 with Cu in a d^9 state led to an electronic structure that could be modeled as a half-filled, single-band Hubbard model

$$H = -\sum_{\langle ij \rangle \sigma} t_{ij} c_{i\sigma}^\dagger c_{j\sigma} + U \sum_i n_{i\uparrow} n_{i\downarrow} \tag{1}$$

on a 2D square lattice of the copper-oxygen planes. He argued that these materials are in the large U limit, with on-site Coulomb repulsion U much larger than the kinetic energy t, so that the half-filled system is a Mott insulator. The cuprates are unique examples of 2D $S=1/2$ single-band Mott insulators that can be doped with carriers. Thus the problem of high-T_c superconductivity was one of understanding a lightly doped Mott insulator.

This basic picture underlies much of the subsequent work in the field. It is now well established that the undoped parent compound is a Mott insulator (technically, a charge transfer insulator), and that upon doping, the holes live in a single band (one per CuO_2 plane) that is an admixture of Cu $d_{x^2-y^2}$ with oxygen $p\sigma$ orbitals.

The parameters of the 2D Hubbard model that describes the cuprates are by now well known from photoemission and electronic structure calculations. Nearest-neighbor and next nearest-neighbor hoppings are necessary to describe the observed shape of the Fermi surface[5] and $U \sim 10t$. The resulting antiferromagnetic (AFM) superexchange $J \sim t^2/U \sim 100 - 150 \, \text{meV}$ (from neutron experiments). It is important to note that while J is the "smallest" energy scale in the problem ($J < t \ll U$) it is nevertheless very large compared with most other transition metal oxides. We shall see that J sets the scale for pairing in RVB theory.

One must keep in mind that the Hubbard model is a minimal model for any real material, and questions remain about whether interactions beyond the on-site Hubbard U could play an important role. Small perturbations to the Hubbard model could have important consequences for the stability of various phases, given how close in energy various competing states are for a lightly doped Mott insulator. But there is no doubt that understanding the 2D Hubbard model, or its low-energy version, the t-J model, remains one of the central challenges in the field. The Hubbard model had long been recognized as one of the simplest lattice model of interacting fermions, occupying a place in many-body theory analogous to that of the Ising model in classical statistical mechanics. We know little with certainty about the Hubbard model away from half-filling, except in some limiting cases (one dimension, infinite dimensions, weak coupling, high-temperatures). Historically it had been invoked to understand antiferromagnetism, ferromagnetism and the metal-insulator transition, but prior to the late 1980s it had not been seriously thought of as a model for superconductivity.

3. The Mott Insulator

What is the ground state of the parent Mott insulator? The 1987 RVB paper suggested that the 2D $S = 1/2$ antiferromagnet on a square lattice might have sufficient quantum fluctuations to destroy Néel order and a spin liquid ground state would be realized. The RVB idea was, in fact, first introduced by Anderson and Fazekas[6] many years earlier, to describe a quantum spin liquid as a "resonating" linear superposition of "valence bonds" or singlet pairs of electrons.

Very soon after the discovery of HTSC, however, it became evident — both from neutron experiments and from theoretical studies — that La_2CuO_4 had a Néel ordered ground state. This might seem reason enough to dismiss the entire RVB approach. However, as detailed below, the RVB framework provides a very powerful way to think about the *doped* materials, despite the fact that the undoped system is not a spin liquid. (I note in passing that the Néel ordered state can also be described in RVB language with a suitable distribution of "bond lengths".[7])

There has been tremendous progress on understanding quantum spin liquids in recent years. On the experimental front, certain triangular lattice organics and kagome lattice oxides have emerged as compelling candidates for spin liquid ground states. Theoretically there has been much progress in understanding their properties using numerical tools like DMRG and new ideas of topological order, deconfined gauge theories and entanglement entropy. But I must return to the cuprates.

4. Are Correlations Important in the Doped System?

Central to the 1987 RVB paper was the large Coulomb U. In the infinite U limit, the Hilbert space at each site is reduced from four states to three $|\uparrow\rangle$, $|\downarrow\rangle$, and $|0\rangle$ (or a vacancy) for the hole-doped case (density $\langle n\rangle \leq$ 1 particle per site). The fourth state $|\uparrow\downarrow\rangle$ with double occupancy can be projected out by the Gutzwiller projection operator $\mathcal{P} = \prod_i (1 - n_{i\uparrow}n_{i\downarrow})$. The effects of a large but finite U can then be taken into account[8] through a canonical transformation e^{-iS} calculated perturbatively in t/U. This leads to appearance of the crucial superexchange scale $J = 4t^2/U$. It is precisely this transformation acting on the Hubbard Hamiltonian $e^{iS}He^{-iS}$ that leads to the tJ Hamiltonian, which then operates on the projected, low-energy Hilbert space.

While the large U is clearly important at half-filling and leads to the Mott insulator, how do we know that it continues to be important in the doped materials? Could it not be that screening makes the effective Coulomb interactions weak and then they can be treated using conventional diagrammatic methods,[9,10] rather than the non-perturbative projection $\mathcal{P}$?

Perhaps the most direct — and model-independent — evidence for the strong-coupling point of view is the large *particle-hole asymmetry*,[11–13] with a characteristic doping dependence, seen in all STM (scanning tunneling microscopy) spectra.[14] One can derive exact sum rules[12] for the single particle spectral function of Gutzwiller projected fermions, without making any assumptions about the nature of the ground state or translational invariance. The integrated spectral weight $S^-(\mathbf{r})$ for *extracting* an electron at a point $\mathbf{r}$ is proportional to the local density $n(\mathbf{r}) \equiv 1 - x(\mathbf{r})$, where $x(\mathbf{r})$ is the local hole doping. Thus $S^-(\mathbf{r}) \equiv \int_{-\infty}^0 d\omega N(\mathbf{r}, \omega) = 1 - x(\mathbf{r})$ where $N(\mathbf{r}, \omega)$ is the local DOS measured in STM. On the other hand the integrated spectral weight for *injecting* an electron is $S^+(\mathbf{r}) \equiv \int_0^\Omega d\omega N(\mathbf{r}, \omega) = 2x(\mathbf{r}) + \mathcal{O}(t/U)$ where the two arises from spin and the upper cutoff Ω (with $t, J \ll \Omega \ll U$) is the result of the projection that excludes the upper Hubbard band. Thus the spectral weight for extracting an electron S^- is much larger that the low-energy spectral weight to add an electron S^+ in a doped Mott insulator, and the asymmetry S^-/S^+ increases with underdoping $x \to 0$. This doping-dependent asymmetry between the positive and negative bias conductance is manifestly visible in all STM data.[14] The factor of two in S^+ means that upon doping a Mott insulator, one does not just change the chemical potential in a "rigid band structure" but rather there must be a transfer of spectral weight from the upper Hubbard band to the low energy subspace, where new state are created.

5. Superconductivity in Doped Mott Insulators

The RVB approach to superconductivity (SC) differs from the standard BCS picture in several ways. (1) The SC arises from inherently repulsive inter-actions, with the superexchange J responsible for pairing. (2) The SC state is best thought of in real-space as a soup of singlet pairs with Gutzwiller projection removing all configurations with double occupancy. (3) SC does not arise as a Fermi surface instability, but rather as an instability of the Mott insulator. To quote from the 1987 RVB paper "the pre-existing mag-netic singlet pairs of the insulating state become charged superconducting pairs when the insulator is doped."

It was recognized quickly that the RVB approach unequivocally implied[15–18] a d-wave pairing state rather than the s-wave pairing discussed in the very early papers.[2,19] The on-site node in the d-wave pair wave function is naturally energetically favorable in the presence of a large Coulomb U. In fact, d-wave pairing was predicted earlier even for the weak coupling Hubbard model.[9] Despite these clear theoretical predictions, there was a strong resis-tance to accepting the idea of d-wave pairing in the cuprates, in large part due to their observed insensitivity to disorder. In any case by the mid-1990s it was established beyond doubt, largely as a result of Josephson interferom-etry experiments,[20] that the cuprates were in fact d-wave superconductors. I will return below to the very important question of how a d-wave SC can be so robust against disorder?

For a variety of reasons, including the absence of the quantum spin liquid at half-filling, RVB ideas largely fell out of favor through the 1990s, except for valiant attempts to understand the role of gauge fluctuations about the RVB slave boson mean-field solution.[21] Anderson himself was exploring other approaches to HTSC — including interlayer tunneling theory and ways of understanding the breakdown of Fermi liquid theory in the normal state — that are described in his book,[22] which has remarkably little discussion of RVB.

RVB redux: In 2001, Paramekanti, Trivedi and I resurrected the original RVB idea for the superconducting state.[23] (Phil later dubbed this the "plain vanilla" version[24] of RVB.). Even if there was no quantum spin liquid at half-filling, the natural description of superconductivity in a doped Mott insulator is in terms of a Gutzwiller projected BCS wave function. This is an RVB state, a linear superposition of singlet pairs with holes or vacancies, which permits the pairs to be mobile, indeed superconduct. The size and internal (d-wave) structure of the pairs is determined by the variational principle, and one can ask: How do strong correlations affect the properties of the SC state? What insights do we get into observable properties using this variational approach to the large-U Hubbard model in 2D? The time was ripe for RVB to confront experiments.

I first briefly describe the insights from the variational calculation of SC state, their surprising ability to make contact with experiments despite the simplicity of the theory, and also discuss more recent theoretical and experimental developments that show the need to go beyond this simple theory.

The no-double-occupancy constraint of $\mathcal{P}$ is implemented either using variational Monte Carlo (exact but numerical) or using the Gutzwiller approximation (which is analytical). The two methods give qualitatively (often, semi-quantitatively) similar results.[25,26] The variational pairing gap parameter $\Delta(x)$ is found to be a monotonically decreasing (almost linear) function of hole density x but the SC order parameter $\Phi(x)$ shows a non-monotonic dome as a function of x. This is in marked contrast to ordinary BCS theory, where the two would have just been proportional to each other. The order parameter dome — which is a surrogate for T_c in this ground state calculation — is a natural consequence of projection $\mathcal{P}$. Strong correlations suppress number fluctuations as one approaches the Mott insulator and thus enhance quantum phase fluctuations which kill superconductivity as $x \to 0$. In fact, the same suppression is also seen in the $T = 0$ superfluid density $\rho_s(x)$ in this theory, and is in qualitative agreement with experiments.[27] I emphasize that the primary reason for the suppression of SC on the underdoped side of the dome comes from the largest energy scale in the problem U (i.e., approaching the Mott insulator), and *not* a competing order parameter, whose characteristic scale can at most be of order J. Of course, competing order parameters — discussed further below — can lead to a *further* suppression of SC, for instance stripe formation is known to lead to a deep hole in the T_c dome in certain La-based materials and CDWs are the cause of the T_c plateau in YBCO.

The projected wave functions also lead to detailed predictions for the low-energy excitations in the SC state, the nodal quasiparticles. These excitations have a quasiparticle weight Z and a spectrum $E_{\mathbf{k}} \approx \sqrt{v_F k_\perp^2 + v_\Delta k_\parallel^2}$ near the node, where v_F is the Fermi velocity and v_Δ the slope of the SC gap at the node. All of these quantities were found to have very interesting doping dependences in projected SC state. $Z(x) \simeq 2x/(1+x)$ signaling loss of single-particle coherence as one goes toward the Mott insulator at $x = 0$. $v_F(x)$ was found to be renormalized by a factor of 2 or 3 below its bare value but with very weak x-dependence. This unusual result — vanishing Z but constant m^* — arises from an interesting interplay of the $\mathbf{k}$- and ω-dependences of the self-energy. $v_\Delta(x)$ was found to track the variational $\Delta(x)$ and thus found to be increasing as $x \to 0$.

To what extent do these predictions agree with experiments? The $Z(x)$ prediction is in excellent agreement with angle-resolved photoemission (ARPES) measurements.[28] The constant $v_F(x)$ was in excellent agreement with ARPES experiments that followed soon after the prediction, but recent high resolution ARPES data[29] that probes the spectrum within few meV of the chemical potential shows that at low energies $v_F(x)$ gets suppressed as SC is destroyed with underdoping. The gap slope $v_\Delta(x)$ does grow as x decreases starting from the overdoped side, but the latest ARPES experiments show that it saturates[29] at about 40 meV over a broad doping range, even while the maximum (antinodal) gap is increasing as x decreases. STM data are broadly consistent with this but involve the combination $N(\omega) \sim Z|\omega|/v_F v_\Delta$, while thermal transport probes v_F/v_Δ. The qualitative difference between the doping dependence of v_Δ and the antinodal gap — often called "two gap" behavior — remains to be understood. A possible explanation might be a competing order like, e.g., charge density waves (CDW). But if this order was strong enough to open up a large gap at the antinode, it would also be expected to impact the electronic dispersion in other ways (shadow bands, pockets, gaps below the chemical potential) that are inconsistent[30] with the ARPES data.

Competing orders: I next discuss states that compete with d-wave SC: long range AFM order (which exists in all cuprates at half-filling) and stripes (which exist in some materials in the underdoped regime). I will return to CDW below when I discuss the pseudogap and the underdoped materials in high magnetic fields.

The projected wave function approach shows a first-order transition at a finite doping leading to a coexistence[31] of AFM and d-wave SC before SC disappears altogether to lead to an AFM Mott insulator at half-filling. The same analysis also gives insights (a) into the asymmetry of hole-doped and electron-doped HTSC, with a much larger regime of AFM in the latter relative to SC, and (b) into the relationship[32] between SC and the range of hopping parameters in the CuO_2 plane.

Another important question concerns any variational analysis for a Hamiltonian with multiple states that are very close in energy density. This has certainly been known to be the case for the large-U Hubbard and t-J models, and is illustrated in recent results obtained tensor network techniques,[33] which show that stripes can be energetically competitive with d-wave SC in the t-J model. In short, small perturbations to the Hamiltonian can led to drastic changes in the answers. Nevertheless, since SC is the dominant broken symmetry in HTSC's at zero field and $T = 0$, the projected SC wave function approach is very useful to gain insights into cuprates.

Disorder: The insensitivity of the d-wave cuprates to impurities is puzzling, given that disorder is a pair-breaking perturbation according to the conventional theory of dirty SCs. There are three reasons why the cuprates are robust against disorder. First, the dopants, an intrinsic source of disorder, are far from the CuO_2 planes. Second, even when the CuO_2 planes are intentionally disordered (say, by Zn substitution for Cu) the short coherence length gives rise to a spatially inhomogeneous response, with a strong local suppression of SC without completely destroying it globally. Finally, there is a more subtle reason: correlations protect the system against disorder.

Combining the Gutzwiller approximation with an inhomogeneous Bogoliubov deGennes analysis, we take into account the combined effects of correlations and disorder in a short coherence length SC. We find[34] that correlations have a dramatic effect in protecting the low energy excitations near the nodes of the d-wave gap, which are much less impacted by impurity scattering compared with the higher energy excitations near the antinode. This nodal-antinodal dichotomy is directly visible in both STM[35] and ARPES[5,36] experiments. While the theoretical results of Ref. 34 give very interesting insights, a deeper understanding of *why* correlations protect the d-wave SC against disorder remains elusive at the moment.

6. Pseudogap

The abnormal "normal" states of the HTSC have proved to be much harder to understand than the broken symmetry states, which are are at least adiabatically connected to states where we can control calculations. The normal states on the other hand have no order parameter and no sharp quasiparticles, and seem to differ qualitatively from the Landau Fermi liquid normal state of conventional metals.

The pseudogap regime lies in a range of doping values roughly between the optimal hole doping with the highest SC T_c and the AFM Mott insulator. The pseudogap anomalies are observed in the temperature range $T_c < T < T^*$, where $T^*(x)$ is (most likely) a crossover temperature that increases with decreasing hole doping and has roughly the same

x-dependence as the large antinodal gap. These anomalies[21,37] all imply a severe loss of low-energy spectral weight with an onset near T^* as seen in the spin susceptibility, the NMR relaxation rate, specific heat, c-axis optical conductivity, STM and ARPES. ARPES,[5] in particular, has led to detailed insights into the anisotropy of the pseudogap and how it leads to truncated "Fermi arcs."[38] The arc is a disconnected segment in $\mathbf{k}$-space, centered about the zone diagonal, on which the spectral function peaks at the chemical potential. Beyond the arc, near the zone boundary, the spectral function shows a large suppression at low energies — the pseudogap.

Early on, it was suggested that some aspects of the pseudogap could be understood in terms of the existence of pairing without phase coherence[39,40] above T_c. But THz,[41] Nernst, diamagnetism[42] and STM[43] experiments all show that although the onset of "SC fluctuations" occurs well above above T_c in the underdoped regime, such effects do not persist all the way up to T^*. Further the x-dependence of this onset is non-monotonic like $T_c(x)$ unlike the pseudogap temperature $T^*(x)$. Thus although spin-pairing persists up to T^*, as evidenced by the NMR anomalies, "SC fluctuations" do not.

Recent X-ray measurements[44,45] have clearly revealed the importance of CDWs in underdoped cuprates. In the zero field pseudogap state, the CDW correlation length grows to be as large as 15–20 lattice spacings just above the SC T_c, and then the onset of SC long-range order suppresses the CDW. As a magnetic field is turned on perpendicular to the CuO_2 plane, the CDW grows as seen in the X-rays, NMR and also in STM data where one sees CDWs nucleated around vortex cores.[47] There are also intriguing observations of broken time-reversal in many underdoped cuprates, possibly arising from intra-unit cell current loops.[46]

One of the most remarkable developments has been the observation of quantum oscillations[48] once the magnetic field has destroyed SC long range order and the system has entered a resistive state. The first surprise here is that different ways of destroying SC lead to very different states. If SC is destroyed at $H = 0$ with $T > T_c$, one enters the non-Fermi liquid pseudogap state. However, if SC is destroyed by turning on a large field, the low temperature state is a Fermi liquid that exhibits quantum oscillations. The second surprise is that the oscillation frequency implies a Fermi surface (FS) area that is impossible to reconcile with the doping level. The only resolution is that the FS is reconstructed by a (field-induced) broken translational symmetry like a CDW. While the majority view is that the field-induced resistive state with CDW order is just a simple Fermi liquid, there is an intriguing observation of quantum oscillations in the specific heat[49] that shows a $\sqrt{H}$

background persisting into the resistive state, and may be the signature of a d-wave vortex liquid.[50]

7. The Strange Metal

The "strange metal" normal state exists in a fan-shape region of the cuprate phase diagram above optimal doping. It is characterized by anomalous transport — including the famous linear-T resistivity — and ill-defined electronic excitations. It nevertheless has a locus of gapless excitations that looks like a conventional Fermi surface enclosing $(1 + x)$ holes, despite the absence of sharp quasiparticles.

The early RVB mean field theories[21] described this region as one in which the holons are not condensed and the spinons are not paired. This may seem to be a reasonable zeroth order description, but it is neither under control theoretically nor is it able to explain the observed marginal Fermi liquid phenomenology.[51] It may also be tempting to describe this regime as a quantum critical fan, where the only energy scale is T, but it is not clear which quantum critical point, if any, controls this behavior, and why there seems to be no observable divergent length scale associated with a vanishing energy scale. Anderson has described the strange metal behavior in terms of a hidden Fermi liquid,[52] and recent calculations of projected fermions[53] and DMFT studies[54] may provide a microscopic underpinning for these ideas. There is now a growing body of experimental evidence that the linear-T resistivity is a universal feature of the high-temperature incoherent state of many correlated metals, and not just limited to the normal states of HTSCs.

8. Summary

In summary, we do understand many aspects of the strongly correlated SC state in the cuprates. The RVB idea of a Gutzwiller projected superconductor has led to a qualitative, and in some cases semi-quantitative, understanding of a variety of SC state properties including the d-wave pairing, the short coherence length, a superfluid stiffness much smaller than the energy gap at low doping, the qualitatively different behavior of the energy gap and the SC order parameter as a function of doping, the doping dependence of the nodal quasiparticles, and the insensitivity of the SC to disorder.

A comprehensive understanding of the non-superconducting states of the cuprates is still a work in progress. These challenges include the questions of the breakdown of Fermi liquid theory in the strange metal and the pseudogap, and also the emergence of sharp quasiparticles in the high-field state with a broken translational symmetry. (For a recent review with more extensive references see Ref. 55.)

High-temperature superconductivity in the cuprates is surely one of the most fascinating examples of emergence in condensed matter physics. Add a few holes to a 2D $S = 1/2$ single band Mott insulator, and entirely unexpected, new set of phenomena emerge.

9. Concluding Remarks

High-T_c is just one of many fields that Phil has contributed to. The sum total of Phil Anderson's contributions to physics is staggering. He has made pioneering contributions to magnetism (exchange, spin waves, local moments, Kondo problem, spin liquids), to superconductivity and superfluidity (broken symmetry, gauge invariance, Anderson–Higgs mechanism, dirty superconductors, Josephson effect, flux creep, superfluid ^{3}He, high-T_c) to disordered systems (localization, spin glass, glasses), and to the ideas of emergence and complexity.

About a dozen years ago, I had occasion to address an audience that included a large group of UNESCO ambassadors at the ICTP in Trieste. I was supposed to tell them about condensed matter physics in general and — for reasons that are not relevant here — about P. W. Anderson, in particular. At first, I just could not see how to communicate the breadth and depth of the contributions listed above to a group that probably knew more about the C (culture) in UNESCO than the S (science). So I decided to show them a slide with several paintings of Pablo Picasso taken from various periods of *his* long career. I said that if we did not know that all of these paintings were made by the same artist, we would be hard pressed to believe it. The same is true of Phil's contributions to physics. But Nandini Trivedi pointed out to me that there was another, more profound, similarity between the two. In his famous sequence of drawings of the bull, Picasso does a "renormalization group transformation" in front of our eyes and arrives at the "fixed point Hamiltonian" describing the essence of the beast. This is precisely what Phil has done so often for our field, and why there is a different Anderson model or Hamiltonian in different areas of condensed matter physics.

Acknowledgments

I would like to acknowledge the support of the NSF and the DOE in my research on high-T_c superconductors, most recently through grants NSF-DMR-1410364 and DOE-BES Grant No. DESC0005035.

References

1. P.W. Anderson, *Science* **177**, 393 (1972).
2. P.W. Anderson, *Science* **235**, 1196 (1987).
3. J.C. Bednorz and K.A. Muller, *Z. Phys. B* **64**, 189 (1986).
4. M.K. Wu *et al.*, *Phys. Rev. Lett.* **58**, 908 (1987).
5. A. Damasceli, Z. Hussain and Z.X. Shen, *Rev. Mod. Phys.* **75**, 473 (2003); J.C. Campuzano, M.R. Norman and M. Randeria, in *Physics of Superconductors*, Vol. II, eds. by K.H. Bennemann and J.B. Ketterson (Springer, Berlin, 2004), p. 167.
6. P.W. Anderson, *Mater. Res. Bull.* **8**, 153 (1973); P. Fazekas and P.W. Anderson, *Philos. Mag.* **30**, 23 (1974).
7. S. Liang, B. Ducot and P.W. Anderson, *Phys. Rev. Lett.* **61**, 365 (1988).
8. C. Gros, R. Joynt and T.M. Rice, *Phys. Rev. B* **36**, 381 (1987).
9. D.J. Scalapino, J.E. Loh and J.E. Hirsch, *Phys. Rev. B* **34**, 8190 (1986); see also: D.J. Scalapino, *Rev. Mod. Phys.* **84**, 1383 (2012).
10. A. Chubukov, D. Pines and J. Schmalian, in *The Physics of Superconductors*, eds. K.H. Benneman and J.B. Ketterson (Springer, Berlin, 2008), p. 1349.
11. M.B.J. Meinders, H. Eskes and G. Sawatzky, *Phys. Rev. B* **48**, 3916 (1993); H. Eskes *et al.*, *Phys. Rev. B* **50**, 17980 (1994).
12. M. Randeria *et al.*, *Phys. Rev. Lett.* **95**, 137001 (2005).
13. P.W. Anderson and N.P. Ong, *J. Phys. Chem. Solids* **67**, 1 (2006).
14. K. McKelroy *et al.*, *Nature* **422**, 592 (2003); H. Hanaguri *et al.*, *Nature* **430**, 1001 (2004).
15. G. Kotliar and J. Liu, *Phys. Rev. B* **38**, 5142 (1988).
16. C. Gros, *Phys. Rev. B* **38**, 931 (1988).
17. Y. Suzumura, Y. Hasegawa and H. Fukuyama, *J. Phys. Soc. Jpn.* **57**, 2778 (1988).
18. F.C. Zhang, C. Gros, T.M. Rice and H. Shiba, *Supercond. Sci. Tech.* **1**, 36 (1988).
19. G. Baskaran, Z. Zhou and P.W. Anderson, *Solid State Commun.* **63**, 973 (1987).
20. D. van Harlingen, *Rev. Mod. Phys.* **67**, 515 (1995); C. Tsuei and J. Kirtley, *Rev. Mod. Phys.* **72**, 969 (2000).
21. P.A. Lee, N. Nagaosa and X.G. Wen, *Rev. Mod. Phys.* **78**, 17 (2006).
22. P.W. Anderson, *The Theory of Superconductivity in the High-T_c Cuprate Superconductors* (Princeton University Press, 1997).
23. A. Paramekanti, M. Randeria and N. Trivedi, *Phys. Rev. Lett.* **87**, 217002 (2001) and *Phys. Rev. B* **70**, 054504 (2004).
24. P.W. Anderson, P.A. Lee, M. Randeria, T.M. Rice, N. Trivedi and F.C. Zhang, *J. Phys.: Condens Matter* **16**, R755 (2004).

25. M. Randeria, R. Sensarma and N. Trivedi, in *The Hubbard Model: Theoretical Methods for Strongly Correlated Systems*, eds. F. Mancini and A. Avella (Springer, 2012).
26. B. Edegger, C. Gros and V.N. Muthukumar, *Adv. Phys.* **56**, 927 (2007).
27. Y.J. Uemura *et al.*, *Phys. Rev. Lett.* **62**, 2317 (1989); M. Broun *et al.*, *Phys. Rev. Lett.* **99**, 237003 (2007); I. Hetel, T.R. Lemberger and M. Randeria, *Nat. Phys.* **3**, 700 (2007).
28. D.L. Feng *et al.*, *Science* **289**, 277 (2000); P.D. Johnson *et al.*, *Phys. Rev. Lett.* **87** 177007 (2001).
29. I. Vishik *et al.*, *Proc. Natl. Acad. Sci.* **109**, 18332 (2012).
30. M.R. Norman *et al.*, *Phys. Rev. B* **76**, 174501 (2007).
31. S. Pathak, V.B. Shenoy, M. Randeria and N. Trivedi, *Phys. Rev. Lett.* **102**, 027002 (2009).
32. E. Pavarini *et al.*, *Phys. Rev. Lett.* **87**, 047003 (2001).
33. P. Corboz, T.M. Rice and M. Troyer, *Phys. Rev. Lett.* **113**, 046402 (2014).
34. A. Garg, M. Randeria and N. Trivedi, *Nat. Phys.* **4**, 762 (2008).
35. K. McElroy *et al.*, *Phys. Rev. Lett.* **94**, 197005 (2005); *Science* **309**, 1048 (2005); M. Vershinin *et al.*, *Science* **303**, 1995 (2004); A.N. Pasupathy *et al.*, *Science* **320**, 196 (2008).
36. X.J. Zhou *et al.*, *Phys. Rev. Lett.* **92**, 187001 (2004); K.M. Shen *et al.*, *Science* **307**, 901 (2005).
37. T. Timusk and B.W. Statt, *Rep. Prog. Phys.* **62**, 61 (1999).
38. A. Kanigel *et al.*, *Nat. Phys.* **2**, 447 (2006).
39. M. Randeria, N. Trivedi, A. Moreo and R.T. Scalettar, *Phys. Rev. Lett.* **69**, 2001 (1992); N. Trivedi and M. Randeria, *Phys. Rev. Lett.* **75**, 312 (1995).
40. V.J. Emery and S.A. Kivelson, *Nature* **374**, 434 (1995).
41. J. Corson *et al.*, *Nature* **398**, 221 (1999).
42. Y. Wang, L. Li and N.P. Ong, *Phys. Rev. B* **73**, 024510 (2006); L. Li *et al.*, *Phys. Rev. B* **81**, 054510 (2010).
43. K. Gomes *et al.*, *Nature* **447**, 569 (2007).
44. G. Ghiringhelli *et al.*, *Science* **337**, 821 (2012); J. Chang *et al.*, *Nat. Phys.* **8**, 871 (2012).
45. R. Comin *et al.*, *Science* **343**, 390 (2014); E.H. da Silva Neto *et al.*, *Science* **343**, 393 (2014).
46. C.M. Varma, *Phys. Rev. B* **55**, 14554 (1997).
47. J.E. Hoffman *et al.*, *Science* **297**, 1148 (2002).
48. N. Doiron-Leyraud *et al.*, *Nature* **447** 565 (2007); S. Sebastian *et al.*, *Nature* **454** 200 (2008).
49. S.C. Riggs *et al.*, *Nat. Phys.* **7**, 332 (2011).
50. S. Banerjee, S. Zhang, and M. Randeria, *Nat. Commun.* **4**, 1700 (2013).
51. C.M. Varma *et al.*, *Phys. Rev. Lett.* **63**, 1996 (1989).
52. P.W. Anderson, *Phys. Rev. B* **78**, 174505 (2008); P.A. Casey and P.W. Anderson, *Phys. Rev. Lett.* **106**, 097002 (2011).
53. B.S. Shastry, *Phys. Rev. Lett.* **107**, 056403 (2011).
54. W. Xu, K. Haule and G. Kotliar, *Phys. Rev. Lett.* **111**, 036401 (2013).
55. B. Keimer, S.A. Kivelson, M.R. Norman, S. Uchida and J. Zaanen, *Nature* **518**, 179 (2015).

Paired Insulators and High-Temperature Superconductors

T. H. Geballe and S. A. Kivelson

Department of Physics, Stanford University
Stanford, CA 94305-4090, USA
geballe@stanford.edu
kivelson@stanford.edu

In common with all condensed matter physicists of our generations, our way of thinking about our field was shaped and greatly inspired by countless seminal works of Phil Anderson — a debt we are pleased to have the opportunity to acknowledge. Discussing plans for this article, we spent many pleasant times debating which particular contribution to highlight — super-exchange (THG), Anderson–Higgs (SAK), the Anderson–Morel pseudopotential (THG), poor man's scaling (SAK), etc. In the end, we opted to highlight a single specific paper which greatly affected each of us at the time, and which has continued to exert a strong intellectual influence on us in the ensuing years. Almost 40 years ago in Ref. 1, Phil introduced the negative U center to account for the fact that most glasses and amorphous semiconductors are diamagnetic. This paper has been highly influential, but certainly does not rank among Phil's most famous works; however, focusing on it enables us to re-acquaint a younger generation with another of Phil's contributions, and to use this as a springboard to discuss some forward-looking extensions that continue to fascinate us.

The inspiration for this work — as with much of Phil's work — was a set of simple experimental facts that make the conclusion almost self-evident when brought into conjunction by the master: Many amorphous semiconductors are highly insulating, even at room temperature, despite the fact that there is direct evidence of a large density of states at the Fermi energy; this indicates that the states at the Fermi energy must be strongly localized. Nonetheless, these materials are often diamagnetic (exhibiting neither Curie nor Pauli paramagnetism), which implies that there must be a "spin-gap" of sorts. These observations can be reconciled, Phil observed, if there is a constant density of localized states with a strongly attractive negative U, so that in equilibrium, each state is either empty or occupied by a singlet pair of electrons. Assuming that the negative U must, in turn, be the consequence of a strong local electron-phonon coupling, he noted that if it is derived from a Holstein model, there is an accompanying exponentially large Franck–Condon reduction of the effective tunneling rate between neighboring localized states, which accounts for the absence of any measurable (hoping) conductance within the band of localized states.

The experiments Phil had in mind included measurements on amorphous and glassy semiconductors. In particular he noted that amorphous silicon is paramagnetic whereas chalcogenide glasses, such as germanium selenide, are diamagnetic. At an intuitive level, he suggested associating the localized states with dangling bonds which are occupied by single electrons in Si (where, presumably, U is positive) and by electron pairs in the chalcogenide glasses, where they are negative U centers.

In an aside in his paper, Phil observed that there is an analogy between the existence of an effective attraction between electrons in a superconductor and the negative U centers in an insulating glass. What is left unsaid is that the resulting spin-gap in the superconductor is measured in degrees, and sets the scale of the superconducting T_c, while in the glasses it is measured in electron volts and is related to the activation energy for conduction. In a conventional superconductor, increasing the strength of the electron-phonon coupling leads to an enhancement of T_c, while in the glass, it leads to an exponential increase of the Franck–Condon factor and hence increasingly strong localization. In the superconductor, the phonon frequency, ω_0, is large compared to the spin-gap and T_c is an increasing function of ω_0, while in the glass, ω_0 is small compared to the spin-gap and the Franck–Condon factor is a decreasing function of ω_0. Still, it is hard not to fantasize that the high pairing scale in the glass could somehow be retained in a related conducting phase, where it would become the superconducting gap scale.

Following this idea and the discovery[2] of superconductivity in PbTe lightly doped with Tl, recent work[3] by one of us and collaborators has uncovered many reasons for believing that the superconductivity in Tl-doped PbTe may be caused by negative U centers: (1) Doping PbTe with other cations results only in a doped semiconductor even at carrier concentrations at which superconductivity occurs on Tl doping. (2) The T_c of $Pb_{1-x}Tl_xTe$ is more than an order of magnitude higher than in other low-density semiconductor-superconductors with similar carrier densities. (3) T_c and a temperature dependence of the resistance consistent with scattering from a charge-Kondo impurity both onset above the same characteristic Tl concentration, $x_c \sim 0.03$. (4) The Hall number is proportional to x for $x < x_c$ and then becomes roughly x independent for larger x. Moreover, the same model has been successful in motivating and accounting for experiments[4] on the superconducting state of In-doped SnTe.

In private correspondence with THG while this work was being carried out, Phil initially expressed interested skepticism. He particularly emphasized the fact that when the negative U results from lattice polarization, the Franck–Condon effect should drastically quench the "charge-Kondo" coupling, i.e. the coherent exchange of electron pairs between the localized centers and the conduction band. (His doubts were largely assuaged when further investigation found a minimum in the low temperature resistance of the sort expected from charge Kondo scattering.) There are two possible ways around this rather fundamental issue: (1) Very high frequency phonons could, in principle, mediate a strong attraction without an accompanying large Franck–Condon suppression of coherence, but this is probably hard to achieve in general, and certainly is not relevant in $Pb_{1-x}Tl_xTe$. (2) If the negative U is largely or entirely a consequence electronic correlations, then there is no reason for a large Franck–Condon suppression.

Indeed, the idea that Tl and In doping could lead to negative U centers comes from the quantum chemistry notion that these are "valence skipping elements." More or less independent of its solid state environment, in crystalline materials in which the nominal valence of Tl is $+2$, the symmetry between different Tl sites is always broken so that half the sites have the effective radius expected for Tl^{+1} and the other half corresponding to Tl^{+3}. With intuition derived from an ionic picture of such solids, this is referred to as "disproportionation." This phenomenon is observed, for instance, in both In and Tl monochalgenides, as well as in AgO in which the divalent magnetic Ag^{+2} disproportionates to form equal concentrations of non-magnetic Ag^{+1} and Ag^{+3}. Disproportionation commonly occurs when the ground state of the cation with the nominal valence would be expected to contain a half-filled s-shell.[7] In the crystal, the disproportionation is thought[5,6] to be driven by the stability of filled shells together with the response of the polarizable lattice.

The notion that in valence skipping elements, it is largely a feature of the electronic structure, rather than the polarizability of the surrounding lattice that is responsible for the negative U is a highly non-trivial extension of the original proposal of Anderson. The significance of this idea, and estimates of the effective U for various elements was discussed by Varma,[6] who in particular proposed that the negative U associated with the valence skipping property of Bi underlies both the charge density wave formation and the mechanism of superconductivity in Pb and K-doped $BaBiO_3$. In particular, Tl with a nominal +2 valence and Bi with a nominal +4 valence have the same half-filled 6S orbital, and hence share the same tendency to disproportionate to produce a mixture of sites with $6S^2$ and $6S^0$ configurations, which in the case of Bi can be thought of as a mixture of Bi^{+3} and Bi^{+5}. While an effective attraction which is a direct and moderately local consequence of purely repulsive microscopic interactions between electrons is somewhat counterintuitive, and its role in producing negative U centers remains controversial,[8] as a point of principle the possibility of such an occurrence can be established from studies of the repulsive Hubbard model on suitable clusters; pair-binding has been shown[9] (by exact diagonalization) to occur in suitable ranges of parameters on the Hubbard square, tetrahedron (where it is particularly strong), cube, and truncated (12 site) tetrahedron.

Abstracting what is important from this history reveals that there are two essential conditions needed to turn a paired (negative U) insulator into a good superconductor: (1) The localized pair states must be resonant with the Fermi energy of the itinerant electron system. (2) The amplitude for coherent tunneling of pairs between the localized (negative U) centers and the itinerant band must be substantial. The first condition can be satisfied by tuning the chemical potential; the latter is more difficult and puts limits on the allowable strength of the electron-phonon coupling. In Tl-doped PbTe, nature is kind to us — the introduction of Tl apparently both produces the negative U centers and dopes the valence band until, for $x > x_c$, the chemical potential is such as to make the Tl^{+1} and Tl^{+3} states degenerate.

However, despite considerable circumstantial evidence, direct evidence of negative U centers in metallic systems is still lacking. One possible way to obtain such direct evidence would be to use a non-superconducting layer with putative negative U centers as the normal junction between two superconductors. If single-particle tunneling is the dominant process, the junction characteristics would be expected to satisfy the Ambegoakar–Baratoff[10] relation, $I_c R = \pi\Delta/2e$, where Δ is the superconducting gap, R is the normal state junction resistance, and I_c is the critical current. However, if the critical current reflects resonant tunneling through negative U centers, $I_c R$ can exceed this value by an arbitrarily large factor.[11] Perhaps an experiment to test this idea could be undertaken in Tl-doped PbTl in structures in which the Tl concentration is modulated to define an SNS junction.

In 1987, in one of the two most highly cited papers[12] of his entire stellar career, Phil proposed that high-temperature superconductivity can arise by smooth evolution with doping from a novel "spin-liquid" or "RVB" insulating phase. This idea is broadly related to the notions we have discussed up until now, in the sense that it envisages pairing to be a property of the insulating state which is inherited by the conducting state attained upon doping.[13]

There are, however, many fundamental differences between an RVB insulator and a negative U center insulator; they are, in fact, distinct phases of matter. One practical difference is that the "pairing" in the RVB state is collective, so that there is no issue of an associated Franck–Condon effect suppressing coherence as in the case of localized negative U centers. Indeed, in the original proposal of Anderson, there was no obvious energy scale associated with pairing, but in subsequent work based on the same notion,[14–17] the idea emerged that in the insulating state, the pairing scale corresponds to a spin-gap or spin pseudo-gap, and that this evolves directly into the superconducting gap upon doping. In other words, the spin liquid can be thought of as a superconducting state with vanishing superfluid density,[14–19] and the pairing is more BCS-like than real-space.

The evidence that this family of ideas applies to the cuprates has been summarized in Ref. 20. The idea that superconducting correlations can exist in an insulating phase, and that this can lead to an anomalously large pairing scale is appealing in a more general context. Since the superconducting T_c is determined by the larger of the pairing scale and the scale set by the superfluid stiffness, in any superconductor to which these ideas apply it is likely[19] that phase ordering plays an unusually important role in determining T_c, especially upon close approach to a putative nearly superconducting insulator. A version of such a mechanism[21] is the "spin-gap proximity effect," which is in a sense half-way between a negative U and an RVB scenario. Here, a strongly correlated insulator (of which the two-leg Hubbard ladder can be taken as the paradigmatic example), which is not necessarily in an exotic phase of matter such as a spin-liquid, but which has significant local superconducting correlations, is placed in contact with an itinerant metallic system; then, by an interaction that is formally equivalent to the proximity effect in superconducting-normal metal junctions, assuming only that the chemical potential of the two systems is such as to allow resonant tunneling of electron pairs between them, the spin-gap of the insulator gets transferred to the metal, where it becomes the superconducting gap.

A natural question is whether any of this offers guidance for new directions in the search for novel superconductors. One of us (THG) has long felt that O vacancies may serve as negative U centers in a variety of transition-metal oxides (possibly including certain cuprates), and thus might be a key ingredient in obtaining new superconductors with high superconducting transition temperatures. An example may be offered by SrTiO$_3$,[22] which is superconducting at a substantially lower concentration of carriers, 5×10^{17} cm^{-3}, than any other known superconductor. The Fermi energy at this low carrier density is approximately $E_F \sim 13$K, so the observed superconducting transition temperature, $T_c \sim 0.09$K, can be thought of as being moderately high, $T_c/E_F \sim 10^{-2}$. A plausible case can be made that attributes this high-T_c to oxygen vacancies. Notably, SrTiO$_3$-doped with other n-type dopants (such as Nb) at such low concentrations, is not superconducting. The art of manipulating O vacancies in interesting materials is in its infancy, but if it offers a route to new superconductors, it is certainly worth pursuing.

Concerning this paper, Phil, we hope U is positive.

References

1. P.W. Anderson, *Phys. Rev. Lett.* **34**, 953 (1975).
2. I.A. Chernik and S.N. Lykov, *Fiz. Tverd. Tela* **23**, 3548 (1981) [*Sov. Phys. Solid State* **23**, 2062 (1981)]; S.A. Nemo and Y.I. Ravich, *Phys. Usp.* **41**, 735 (1998).
3. Y. Matsushita, H. Bluhm, T.H. Geballe and I.R. Fisher, *Phys. Rev. Lett.* **94**, 157002 (2005); S. Erickson, J.-H. Chu, M.F. Toney, T.H. Geballe and I.R. Fisher, *Phys. Rev. B* **79**, 024520 (2009).
4. A. Erickson, N.P. Breznay, E.A. Nowadnick, T.H. Geballe and I.R. Fisher, *Phys. Rev. B* **81**, 134521 (2010).
5. B.Y. Moyzhes and A. Drabkin, *Sov. Phys. Sol. State* **25**, 7 (1983).
6. C.M. Varma, *Phys. Rev. Lett.* **61**, 2713 (1988).
7. M.B. Robin and P. Day, Mixed Valence Chemistry, *Advances in Inorganic Chemistry and Radiochemistry* **10**, 247 (1967).
8. W.A. Harrison, *Phys. Rev. B* **74**, 245128 (2006).
9. S.R. White, S. Chakravarty, M.P. Gelfand and S.A. Kivelson, *Phys. Rev. B* **45**, 5062 (1992).
10. V. Ambegaokar and A. Baratoff, *Phys. Rev. Lett.* **10**, 486 (1963).
11. V. Oganesyan, S. Kivelson, T. Geballe and B. Moyzhes, *Phys. Rev. B* **65**, 172504 (2002).
12. P.W. Anderson, *Science* **235**, 1196 (1987).
13. One of us (SAK) remembers hearing Phil talk about this idea in January of 1987, and getting so excited that he declared to his wife, "I know what I am going to be working on for the next year." It is now 27 years later and it still seems that the solution is just around the bend.
14. S.A. Kivelson, D.S. Rokhsar and J.P. Sethna, *Phys. Rev. B* **35**, 8865 (1987).
15. C. Gros, R. Joynt and T.M. Rice, *Z. Phys. B: Condens. Matter* **68**, 425 (1987).
16. G. Baskaran, Z. Zou and P. W. Anderson, *Solid State Commun.* **63**, 973 (1987).
17. G. Kotliar, *Phys. Rev. B* **37**, 3664 (1988).

18. L. Balents, M.P.A. Fisher and C. Nayak, *Int. J. Mod. Phys. B* **12**, 1033 (1998).
19. V.J. Emery and S.A. Kivelson, *Nature* **374**, 434 (1994).
20. P.W. Anderson, P.A. Lee, M. Randeria, T.M. Rice, N. Trivedi and F.C. Zhang, *J. Phys.: Condens. Matter* **16**, R755 (2004).
21. V.J. Emery, S.A. Kivelson and O. Zachar, *Phys. Rev. B* **56**, 6120 (1997).
22. X. Lin, Z. Zhu, B. Fauqué and K. Behnia, *Phys. Rev. X* **3**, 021002 (2013).

Special Properties of High-T_c Cuprates, Radically Different from Other Transition Metal Oxides

T. M. Rice

Institute for Theoretical Physics
ETH Zürich, 8093 Zurich, Switzerland
rice@itp.phys.ethz.ch

The sensational discovery of high-temperature superconductivity in hole doped cuprates was immediately followed by Phil Anderson's insight that this was a radically new form of superconductivity arising in a single band Hubbard model on a simple square lattice doped slightly away from half-filling. Later many more special properties of the cuprates were uncovered. Today we are still some way from a comprehensive description of these anomalies and even further from a consensus on the underlying physics. Their behavior contrasts strongly with other transition metal oxides, where strong electron-phonon coupling leads to very different properties.

1. Cambridge in the Early 1960s and Bell Labs Later in the 1960s and 1970s

My first contact with Phil Anderson came when he spent a sabbatical year at Cambridge in 1961–1962 and I was a graduate student working with Volker Heine in the Solid State Theory group. I was located on the top floor of the Cavendish Lab. and the experimentalists were located in the basement, which hindered interaction between us. It was at this time that Brian Josephson, who was a graduate student with Brian Pippard in the experimental group, developed his famous theory of superconducting tunneling with strong support from Phil Anderson. Also at that time, Brian Pippard was predicting a decidedly gloomy future for solid state physics, enunciated in his infamous 1961 article in *Physics Today*[1] entitled "The Cat and the Cream." The title emphasized the view that his generation had solved all the interesting problems in the field and future generations would be left with uninteresting problems and minutiae. He had reasons for this pessimism. BCS had solved superconductivity a few years earlier, he and David Shoenberg were measuring the Fermi surfaces of elemental metals through the anomalous skin effect and quantum oscillations in magnetic fields, and pseudopotential theory was calculating the band structure of silicon and other semiconductors, etc. So the great challenges that existed at the start of his scientific career were understood. In this atmosphere it was very reassuring to me as a young theorist to hear a much more optimistic view from Phil Anderson. I remember well his seminar in an evening series organized by the graduate students. His lecture contained an early version of his strongly held views that the physics of solids and liquids form a scientific discipline, requiring its own concepts and ideas. These views later became famous, expressed in the article "More is Different".[2] He also expressed his skepticism at that time about the thesis that John Slater was espousing, that in the future all would be revealed through the rapid progress in computers. Of course computational physics has made enormous progress since those days when Fortran was in its infancy, but we still need to analyze exactly what computer simulations and, of course, experiments tell us.

My next contact came when I joined the theory group at Bell Labs in the mid-1960s. As a young inexperienced theorist, I was naturally nervous joining this distinguished group. At that time the Kondo problem was the big open issue in theory and it was strongly represented at Bell Labs, e.g., by Phil Anderson, Jun Kondo, Don Hamann, Joel Appelbaum and others. I quickly decided that if I was going to survive at Bell Labs, I must work in a different area.

Denis McWhan had a new high pressure rig and was using to investigate metal-insulator transitions, a topic that I had gotten involved in during my postdoc years at La Jolla with Walter Kohn. This led us Denis and I to look at the temperature-pressure phase diagrams of transition metal oxides, mainly vanadium oxides. Our biggest achievement was the T-P phase diagram of V_2O_3 and its alloys and the interpretation of it as an example of the elusive Mott transition in an ordered solid.[3] The enhancement of mass in the metal as the transition was approached, triggered Bill Brinkman and me to look into how the Fermi surface disappears as the metallic state transformed into a Mott insulator.[4] We found in a simple Gutzwiller approximation the metallic Fermi surface just faded away through a suppression of the quasiparticle weight and an enhancement of quasiparticle mass. This work was the start of my continuing fascination with strongly correlated transition metal oxides.

These oxides, at the beginning of the transition metal series, are generally characterized by the singlet dimer lattices in the case of d^1 oxides, e.g., VO_2 and antiferromagnetic order for d^2 and higher oxides, e.g., V_2O_3. The former take maximum advantage of the strong binding of a singlet by distorting the lattice to enhance the exchange interaction. This distortion in turn traps carriers that break up the singlets, leading at best to low mobilities and of course no superconductivity. Electron doping leads to the Magneli phases, V_nO_{2n-1} which interpolate between d^1 (VO_2) and d^2 (V_2O_3).[5] These intermediate phases are characterized by low symmetry lattice structures and local charge ordering with charge disproportionated d^1 and d^2 sites. Again, the result is transitions to low symmetry insulators that freeze in this local charge ordering as the temperature is lowered. The result is unfavorable conditions for a quantum mechanically ordered states, e.g., superconductors, as opposed to classically ordered electron structures, e.g., charge disproportionated and bond ordered states. So during this time, the late sixties and early seventies my interest in these transition metal oxides concentrated on these novel metal-insulator transitions. I found them very interesting but their properties were not enough to attract Phil Anderson's active interest. These early transition metal oxides left me with the strong conviction that transition metal oxides were strongly correlated electron fluids, but not at all promising materials for superconductivity.

2. The Discovery of the High-T_c Cuprate Superconductors

The discovery of superconductivity by Bednorz and Mueller[6] in $La_{1-x}Ba_xCuO_4$ at a temperature in the upper 30s came as a great surprise, not least to me. Actually I had a conversation with Alex Mueller at a small meeting in Switzerland, shortly before their paper. He mentioned he was looking for superconductivity in transition metal oxides. I must confess my reaction was rather discouraging!

The initial paper by Bednorz and Mueller reported superconductivity with a transition temperature in the upper thirties, which although a substantial increase on the maximum known value in the low twenties for T_c, it did not seem to me to be out of reach for the standard electron-phonon mechanism. Phil Anderson reacted immediately and very differently. He recognized that this was a very different type of superconductivity and immediately proposed the low energy physics was described by a single band Hubbard model, lightly doped away from a Mott insulating state with antiferromagnetically coupled $S = 1/2$ spins.[7] He went much further and in a matter of weeks after learning about superconductivity, he demonstrated the close relationship between a pairing wave function of the BCS form and a resonant valence bond (RVB) state, which would result from doping a magnetic state formed by a quantum mechanical superposition of different configurations of singlet spin pairs in a Heisenberg AF model. He proposed this as the physical model underlying the cuprate superconductors. Some years earlier he had suggested the possibility of a short range ordered spin liquid formed from quantum mechanically delocalized singlet pairs in a frustrated 2-dimensional lattice. A spin singlet has an energy gain that is 3-times that of a pair with fixed antiparallel spins. But mostly they occur as dimer lattices with local lattice distortions, which enhance their energy. In the cuprates the spins of the Cu^{2+} ions sit on a simple unfrustrated square lattice with long range AF order, but also with strong quantum fluctuations into a local singlet pair configurations. In his famous RVB paper, Phil Anderson writes the pre-existing magnetic singlet pairs of the insulating state become charged superconducting pairs when the insulator is doped sufficiently strongly. This rapid realization of the novel character of cuprate superconductivity is but one remarkable example of Phil Andersons powerful intuition, his sense for new physics and for the essentials of an unexpected discovery.

It is interesting to compare and contrast the cuprates with the vanadates, which I had concentrated on previously. The parent state of the cuprates has a valence Cu^{2+} (i.e., $3d^9$) with a single hole in the e_g subshell of the $3d$ states. The O-octahedron surrounding the Cu^{2+} ion is elongated out of the CuO_2 layer resulting in a splitting of the pair of e_g states. The hole lies in the uppermost state leading to a single isolated half-filled band — a conclusion that remains valid when the on-site Coulomb repulsion in the Cu $3d$ states is taken into account. This simple high symmetry electronic structure of the cuprate layers contrasts strongly with the complex and low symmetry electronic structure that the $3d^1$ state takes in VO_2. In the latter the different orientation of the $3d$-t_{2g} orbitals leads to less hybridization as well as overlapping bands. This makes the crystal structure more susceptible when doped to distortions associated with the local charge. As a result one finds low symmetry structures associated with fractional valence and generally insulating charge ordered groundstates. One exception is V_7O_{13} with an antiferromagnetic metallic groundstate. Interestingly this oxide shows an AF Quantum Critical Point under pressure, but alas still no superconductivity.[8,9] It seems that the low symmetry crystal structure inhibits the quantum mechanical delocalization of singlets that is the essence of RVB. This demonstrates that weak coupling to the lattice and high symmetry of the crystal structure in the cuprates, are crucial to their exceptional superconductivity, the properties that Phil Anderson emphasized at the outset.

3. A Selective History of High-T_c Cuprate Superconductivity

In the intervening period of more than a quarter century, the theory of cuprate superconductivity has, however, evolved unevenly. In the first decade the emergence of elegant phase sensitive experiments established the unconventional d-wave symmetry of the superconductivity a clear contrast to the s-symmetry in a BCS theory based on an attraction due to coupling to phonons. This important difference demonstrated that the repulsive Coulomb interaction dominates on the low energy scale here, which is not surprising given the proximity to a Mott insulating state. Theoretical understanding on other special features of the cuprates has evolved in a rather chaotic fashion and even today a consensus with broad support has not yet emerged on many issues.

There are a number of reasons for this development. While all cuprates superconductors are variants with a single essential component lightly doped CuO_2 layers, their full structure is quite complex and it has taken some time to master the art of preparing high quality samples. Also the old adage, that one should examine the most perfect materials without disorder to unravel the underlying causes of anomalous behavior, does not work well here, since in almost all cuprates, hole doping is achieved by introducing random acceptors. Further, while powerful new techniques have emerged such as Angle Resolved Photoemission (ARPES) and Scanning Tunneling Microscopy (STM), they are limited to certain cuprates, i.e. those that can be cleaved to give good surfaces with nearby layers representative of the bulk. Unfortunately the best ordered cuprates do not qualify for these important experiments. The wide range of experimental tools available today makes for another complication. Namely, that an unexpected experimental glitch can trigger a claim that it contains the key to understanding all anomalous properties. The real challenge comes, however, in constructing a consistent theory that covers all the anomalous properties of the cuprates. We are dealing with a strongly interacting fermionic fluid a notoriously difficult problem.

One of the biggest controversies surrounds the nature of the transition from a full Fermi surface containing a standard electron count of $1-x$ (x: hole doping away from half-filling) to the parent Mott insulator at $x = 0$. In the latter, the Fermi surface is completely gone, replaced by a charge gap 2 eV. In the former, the so-called overdoped region, a full Fermi surface enclosing $1 - x$ electrons is observed. But there are signs of deviations from standard Landau Fermi Liquid behavior e.g., linear rather than quadratic in T terms in the resistivity. Phil Anderson has argued that already this overdoping region requires a strong coupling theory and he has obtained good fits to the anomalous temperature dependence of the resistivity using a "Hidden Fermi Liquid" theory.[10] An alternative approach based on the temperature and energy dependence of the scattering vertex calculated in a 1-loop FRG approximation also found a linear term but with a much too small coefficient. The 1-loop approximation, however, limits these calculations to moderate strengths for the on-site Coulomb interaction and so is not quantitatively accurate for cuprates.[11]

The underdoped cuprates with doping below a critical value, $x < x_c$ (0.2) show a host of anomalous properties in what is generally referred to as the pseudogap phase. The anomalous features first appeared in an early comparison of the spin susceptibility, determined from the NMR Knight shift, between underdoped and optimally doped $YBa_2Cu_3O_{7-\delta}$.[12,13] Samples with δ values around 0.3, showed a very substantial drop in the spin susceptibility in the normal state at $T > T_c$, which contrasted strongly with the constant value in the normal phase in optimally doped samples with $\delta = 0$. This drop is generally interpreted as evidence for a some form of spin gap since it is too large to be attributed to short range antiferromagnetic order. This points towards some singlet pairing, in line with Anderson's RVB proposal discussed earlier. An early series of ARPES experiments in the normal phase at $T > T_c$ showed that full Fermi surface observed in the overdoped region, evolved to a set of 4 disconnected Fermi arcs centered on the four nodal points $(\pm\pi/2, \pm\pi/2)$.[14] In the antinodal regions, near to $(\pm\pi, 0)$ and $(0, \pm\pi)$ the ARPES spectra of these underdoped cuprates show a finite energy gap. Interestingly each of the 4 arcs do not extend beyond the square surface in **k**-space whose sides connect the antinodal points. A similar set of Fermi arcs was deduced from the interference patterns in recent STM spectra at low voltages and low temperatures in the superconducting state of underdoped samples.[15] These patterns, when interpreted as a consequence of weak disorder scattering, allow the determination of the energy spectra of the coherent Bogoliubov quasiparticles in the superconducting state. These spectra allowed Kohsaka *et al.*[15] to observe the Fermi surface underlying these propagating Bogoliubov quasiparticles. It also consists of arcs, similar to those observed directly in the ARPES experiments.

3.1. *Theoretical proposals to describe the pseudogap phase*

The explanation of these four disconnected Fermi surface pieces, rather than connected closed Fermi curves found in standard metals, is a challenge to theory and various proposals have been put forward to explain it. One line focuses on a Fermi surface reconstruction due to the presence of some kind of superlattice order. This requires a breaking of translational symmetry. The simplest such symmetry breaking would be the presence of charge (CDW) or spin (SDW) wave order. Such superlattice order does appear in some underdoped cuprates, which show static stripes, e.g., $La_{1.875}Ba_{0.125}CuO_4$. The order however is complex than originally realized, with interwined SDW and Phase Modulated pairing in addition to the hole stripes.[16,17] However, static spatially modulated states do not occur in all underdoped cuprates. This is apparent from NMR/NQR experiments, that detect can static modulations in the local environment of specific nuclei. These came up negative in the case of the stoichiometric underdoped cuprate, $YBa_2Cu_4O_8$.[18,19] Recent developments have revived interest in this topic as will be discussed later.

Numerical simulation of strongly interacting fermions is also notoriously difficult. Dynamic Mean Field Theory has been applied, to the 2D Hubbard model. It can treat large values of the local repulsion U, but it is basically a real space method with limited resolution in k-space. This is sufficient to show that differentiation emerges in k-space with signs of the pseudogap in the antinodal but not in the nodal directions.[20,21] Importantly this demonstrates that k-space differentiation is present in the Hubbard model, without translational or rotational symmetry breaking. But as yet this method has not yielded details of the evolution of the Fermi surface in this region.

An alternative approach is to look to systems where a repulsive interaction causes insulating behavior at half-filling even at weak values of the local repulsive without breaking translational symmetry. Thus the two requirements for a Mott insulator are satisfied. The simplest such case is the 2-leg Hubbard ladder at half-filling. This system has been extensively discussed in the literature (for a review see Ref. 22). The ground state is non-degenerate and insulating for all $U > 0$ with finite energy gaps in the spin, single particle and pair channels. Further translational symmetry along the ladder is preserved. The energy gap to add Cooper pairs, necessary for a charge gap and insulating behavior, is present only at half-filling. It is interesting to trace its origin in the RG analysis at weak coupling. It arises because of the presence of additional elastic Umklapp scattering processes connecting the 4 Fermi wave vectors exactly at half-filling.

Yang, Rice and Zhang[23] sought to generalize this physics to a 2D lattice and introduced a simple ansatz for the coherent part of the single particle propagator to generalize this RVB physics to 2D to this end.

$$G^{\text{RVB}}(\mathbf{k}, \omega) = \frac{g_t(x)}{\omega - \xi(\mathbf{k}) - \Sigma^{\text{RVB}}(\mathbf{k}, \omega)} \tag{1}$$

with a self-energy

$$\Sigma^{\text{RVB}}(\mathbf{k}, \omega) = \Delta_R^2(\mathbf{k})/(\omega + \xi_0(\mathbf{k})). \tag{2}$$

The energies $\xi(\mathbf{k}) = \xi_0(\mathbf{k})$ are band energies with renormalized coefficients and only nearest neighbor hopping terms are retained in the dispersion $\xi_0(\mathbf{k})$ that enters the self-energy. The YRZ ansatz has a pairing form for the self energy, but which crucially has a divergence on a surface spanned by Umklapp scattering processes rather than the Fermi surface as in BCS theory. Umklapp scattering processes enhance the gap and pinning it to this surface gives rise to a charge gap in the spectrum in the antinodal region of the Brillouin zone. Upon doping away from half-filling, the Fermi surface is truncated to a set of closed Fermi contours centered on the nodal directions. These closed contours however look like arcs due to strong variations in the quasiparticle weight along the contour. This combination of short range pairing order associated with a charge and spin gap, has clear connections to Phil Andersons RVB proposal. More details of this approach, including applications to various anomalous properties in the pseudogap phase and a generalization to superconductivity on the ungapped Fermi contours, can be found in a recent review.[24]

Recently H. B. Yang et al.[25] have refined their ARPES spectra of thermally excited states above the Fermi level obtaining the Fermi surface topologies of underdoped BSCCO. They showed that in the pseudogap phase of underdoped samples, a simple extrapolation of the quasiparticle dispersion beyond the maxima leads to Fermi surfaces which are composed of fully enclosed hole pockets as shown in Fig. 1. The spectral weight of these pockets is vanishingly small near the Umklapp surface. The measured area of the pockets is consistent with the hole density. These ARPES results provide strong support to the YRZ ansatz.

The cuprates, however, continue to throw up surprises — a trait that makes the field especially intriguing. Very recently RIXS and related experiments found evidence the pseudogap doping range for a Giant Phonon Anomaly preceding the onset of superconductivity as the temperature is lowered. These reports have given new life to the controversy to the question of translational symmetry breaking in the pseudogap phase. Inelastic X-ray

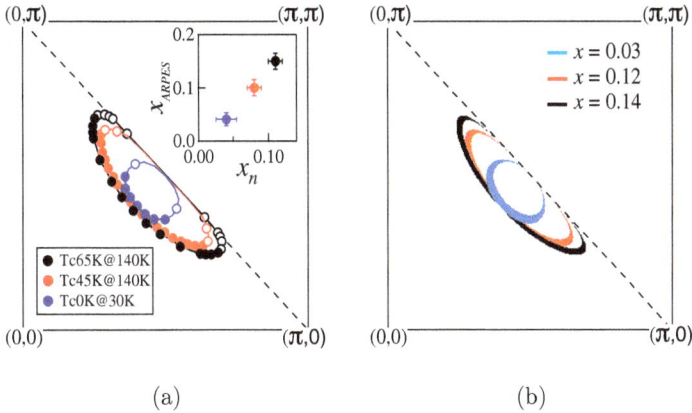

Fig. 1. (a) The pockets determined by ARPES on Bi2212 by H. B. Yang *et al.*[25] for three different doping levels obtained by extrapolating the dispersion beyond the maxima which are observed to lie inside the AF reduced BZ. Inset: The area of the pocket obtained in this way scales with the nominal doping level x_n. (b) For comparison, the Fermi pockets derived from the YRZ ansatz with different doping x.

scattering experiments by LeTacon *et al.*[26] found that phonons in a narrow range around certain wavevectors exhibit extremely large superconductivity-induced line-shape renormalizations. These are accompanied by quasi-elastic "central peaks" due to nanodomains with lattice modulations at these wave vectors. The central question is whether these nanodomains signal a static CDW instability or a disorder driven response to the finite frequency phonon anomalies. LeTacon *et al.*[26] favor the latter interpretation. The absence of a static CDW in the only stoichiometric underdoped cuprate, $YBa_2Cu_4O_8$ in NMR experiments[18,27] supports this interpretation. Later NMR experiments[19] observed low-frequency charge fluctuations accompanying the onset of superconducting fluctuations, which are diminished in the superconducting phase. These results leave us with the puzzle of how the superconducting fluctuations couple so strongly to this particular set of phonons as T_c is approached. Once again, after more than a quarter century of study, the cuprates have demonstrated that our understanding of the pseudogap phase is incomplete and uncovered a fascinating challenge to theory.

I am very grateful to Phil Anderson for very many stimulating discussions and insights. I should also like to thank my many collaborators in the study of strongly correlated electrons stretching over many years, especially, Bill Brinkman, Denis McWhan, Manfred Sigrist, Kai-Yu Yang and Fu Chun Zhang.

References

1. A. Pippard, The cat and the cream, *Phys. Today* **14**(11), 38–41 (2009).
2. P.W. Anderson, More is different, *Science* **177**(4047), 393–396 (1972).
3. D. McWhan, T. Rice and J. Remeika, Mott transition in Cr-doped V_2O_3, *Phys. Rev. Lett.*. **23**, 1384–1387 (1969).
4. W. Brinkman and T. Rice, Application of Gutzwiller's variational method to the metal-insulator transition, *Phys. Rev. B* **2**, 4302–4304 (1970).
5. S. Kachi, K. Kosuge and H. Okinaka, Metal-insulator transition in V_nO_{2n-1}, *J. Solid State Chem.* **6**(2), 258–270 (1973).
6. J.G. Bednorz and K.A. Müller, Possible high-T_c superconductivity in the Ba-La-Cu-O system, *Z. Phys. B: Condens. Matter* **64**(2), 189–193 (1986).
7. P.W. Anderson, The resonating valence bond state in La_2CuO_4 and superconductivity, *Science* **235**(4793), 1196–1198 (1987).
8. P. Canfield, J. Thompson and G. Gruner, Unifying trends found for the V_NO_{2N-1} series by the application of hydrostatic pressure, *Phys. Rev. B* **41**(7), 4850 (1990).
9. H. Ueda, K. Kitazawa, T. Matsumoto and H. Takagi, Charge transport near pressure-induced antiferromagnetic quantum critical point in Magnéli-phase vanadium oxides, *Solid State Commun.* **125**(2), 83–87 (2003).
10. P.A. Casey and P.W. Anderson, Hidden Fermi liquid: Self-consistent theory for the normal state of high-T_c Superconductors, *Phys. Rev. Lett.* **106**(9), 097002 (2011).
11. J.M. Buhmann, M. Ossadnik, T. Rice and M. Sigrist, Numerical study of charge transport of overdoped $La_{2-x}Sr_xCuO_4$ within semiclassical Boltzmann transport theory, *Phys. Rev. B* **87**(3), 035129 (2013).
12. R. Walstedt, W. Warren Jr, R. Bell, R. Cava, G. Espinosa, L. Schneemeyer and J. Waszczak, Cu^{63} NMR shift and linewidth anomalies in the $T_c = 60K$ phase of Y-Ba-Cu-O, *Phys. Rev. B* **41**(13), 9574 (1990).
13. H. Alloul, T. Ohno and P. Mendels, ^{89}Y NMR evidence for a Fermi-liquid behavior in $YBa_2Cu_3O_{6+x}$, *Phys. Rev. Lett.* **63**(16), 1700 (1989).
14. M. Norman, H. Ding, M. Randeria, J. Campuzano, T. Yokoya, T. Takeuchi, T. Takahashi, T. Mochiku, K. Kadowaki, P. Guptasarma *et al.*, Destruction of the Fermi surface in underdoped high-T_c superconductors, *Nature* **392**(6672), 157–160 (1998).
15. Y. Kohsaka, C. Taylor, P. Wahl, A. Schmidt, J. Lee, K. Fujita, J. Alldredge, K. McElroy, J. Lee, H. Eisaki *et al.*, How Cooper pairs vanish approaching the Mott insulator in $Bi_2Sr_2CaCu_2O_{8+\delta}$, *Nature* **454**(7208), 1072–1078 (2008).
16. Q. Li, M. Hücker, G. Gu, A. Tsvelik and J. Tranquada, Two-dimensional superconducting fluctuations in stripe-ordered, $La_{1.875}Ba_{0.125}CuO_4$. *Phys. Rev. Lett.* **99**(6), 067001 (2007).
17. E. Berg, E. Fradkin, E.-A. Kim, S.A. Kivelson, V. Oganesyan, J.M. Tranquada and S. Zhang, Dynamical layer decoupling in a stripe-ordered high-T_c superconductor, *Phys. Rev. Lett.* **99**(12), 127003 (2007).
18. I. Tomeno, T. Machi, K. Tai, N. Koshizuka, S. Kambe, A. Hayashi, Y. Ueda and H. Yasuoka, NMR study of spin dynamics at planar oxygen and copper sites in $YBa_2Cu_4O_8$, *Phys. Rev. B* **49**(21), 15327 (1994).

19. A. Suter, M. Mali, J. Roos and D. Brinkmann, Charge Degree of Freedom and the Single-Spin Fluid Model in YBa$_2$Cu$_4$O$_8$, *Phys. Rev. Lett.* **84**(21), 4938 (2000).

20. S. Sakai, Y. Motome and M. Imada, Doped high-T_c cuprate superconductors elucidated in the light of zeros and poles of the electronic Greens function, *Phys. Rev. B* **82**, 134505, doi: 10.1103/PhysRevB.82.134505. URL http://link.aps.org/doi/10.1103/PhysRevB.82.134505.

21. E. Gull, M. Ferrero, O. Parcollet, A. Georges and A.J. Millis, Momentum-space anisotropy and pseudogaps: A comparative cluster dynamical mean-field analysis of the doping-driven metal-insulator transition in the two-dimensional Hubbard model, *Phys. Rev. B* **82** (15), 155101 (2010).

22. E. Dagotto and T.M. Rice, Surprises on the way from one- to two-dimensional quantum magnets: The ladder materials, *Science* **271** (5249), 618–623 (1996).

23. K.-Y. Yang, T. Rice and F.-C. Zhang, Phenomenological theory of the pseudogap state, *Phys. Rev. B* **73**(17), 174501 (2006).

24. T. M. Rice, K.-Y. Yang and F.-C. Zhang, A phenomenological theory of the anomalous pseudogap phase in underdoped cuprates, *Rep. Prog. Phys.* **75**(1), 016502 (2012).

25. H.-B. Yang, J. Rameau, Z.-H. Pan, G. Gu, P. Johnson, H. Claus, D. Hinks and T. Kidd, Reconstructed Fermi surface of underdoped Bi$_2$Sr$_2$CaCu$_2$O$_{8+\delta}$ cuprate superconductors, *Phys. Rev. Lett.* **107**(4), 047003 (2011).

26. M. Le Tacon, A. Bosak, S. Souliou, G. Dellea, T. Loew, R. Heid, K. Bohnen, G. Ghiringhelli, M. Krisch and B. Keimer, Inelastic X-ray scattering in YBa$_2$Cu$_3$O$_{6.6}$ reveals giant phonon anomalies and elastic central peak due to charge-density-wave formation, *Nat. Phys.* **10**(1), 52–58 (2014).

27. I. Mangelschots, M. Mali, J. Roos, D. Brinkmann, S. Rusiecki, J. Karpinski and E. Kaldis, ^{17}O NMR study in aligned YBa$_2$Cu$_4$O$_8$ powder, *Physica C: Superconductivity* **194**(3), 277–286 (1992).

From Bacteria to Artificial Cells, the Problem of Self-reproduction

Albert Libchaber

Rockefeller University
1230 York Ave, New York 10054, USA
libchbr@rockefeller.edu

Self-reproduction, the production of an offspring identical with the parent, is a fundamental conceptual problem. We first present the historical evolution of the concept "What are structures which produce further identical structures." We then describe the state and development of an artificial cell project, and its feasibility to self-reproduce.

For a naïve physicist like Schrödinger,[1] one of the main questions about life was self-reproduction. He very elegantly exposed the problem. "The physicist and the chemist, investigating inanimate matter, had never witnessed phenomena which they had to interpret in this way. The case did not arise and so our theory does not cover it — our beautiful statistical theory of which we were so justly proud because it allowed us to look behind the curtain, to watch the magnificent order of exact physical laws coming forth from atomic and molecular disorder."

With self-reproduction, we also witness the magnificent order of biological laws coming from atomic and molecular disorder, but it comes as a surprise for the physicist, for it implies a strange world of exponential growth. For example, in a day, with enough feeding material one bacterium *Escherichia coli* will produce about 10^{14} bacteria — a humongous number. We are not used to thinking of a world made of exponentials.

Before developing the problem of cellular reproduction, let us take a historical detour about the history of the cell.

The cell theory started with William Harvey's book about animals generation in 1651.[2] There he proposed that all life forms originate in eggs — "*ex ovo omnia.*" Figure 1 shows that from the egg comes all life forms

Fig. 1.   W. Harvey image: Jupiter opens an egg wherefrom all organisms come out, including humans. All organisms come from eggs. "Ovo omnia" is written on the egg.

including humans, in contradiction with previous strange models of Aristotle and Galen.

The next step came from Robert Hooke who, using a compound microscope, observed cells through a slice of cork. Figure 2 is taken from his book *Micrographia*, published in 1665.[3] There, for the first time, the word "cell" was introduced.

Cells self-reproduction was finally introduced by Robert Virchow in 1858 in his book on cellular pathologie.[4] He sets forth the sentence "*onmia cellula e cellula*," cells come from cells. Figure 3 shows drawings of liver cells presenting nuclear and cell division.

The magnificent order of biological self-reproduction remained as an understood observation for a long time. The first theoretical step came from John von Neumann.[5] It was presented to a group of biologists in 1948 and published in 1951. It was about the general and logic theory of

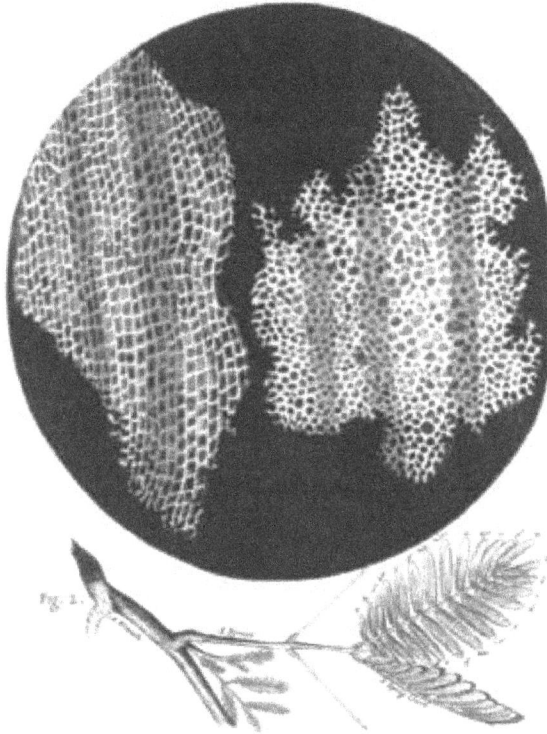

Fig. 2. Cells observed by R. Hooke through a slice of cork with a microscope. A branch of cork is also shown.

automata, more specifically what kind of logical organization is sufficient for an automaton to reproduce itself.

He wished to abstract from the natural self-reproduction problem its logical form. To self-reproduce, any biological machine must follow a sequence of steps: an algorithm. For that, von Neumann turned to Turing universality for computing machines. But "Turing's procedure is too narrow, his automata are purely computing. What is needed is an automaton whose output is another automaton." Figure 4 presents von Neumanns reproduction logic.

"Automaton A which when furnished the description of any other automaton will construct that entity." This is von Neumann's introduction of the Universal Constructor A. In a fascinating way the universal constructor is close to the modern 3D printer. The description itself will be called an instruction I. Automaton B which can make a copy of any instruction I. Combine Automaton A and B with a control mechanism C. Denote the total aggregate by D. To function, D must be furnished with an instruction I_D.

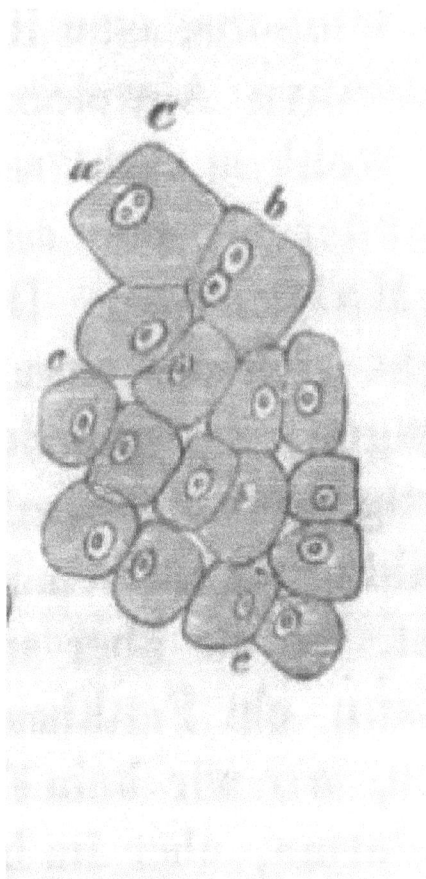

Fig. 3.   Figures of liver cells by R. Virchow showing self-reproduction: (a) A nucleus starts to replicate, (b) the nucleus divides, (c) cells self-reproduce.

The total aggregate is E, as shown in Fig. 4(a). E is self-reproductive. The parts proposed by von Neumann have their counterparts with living cells. Tape I is the DNA genome, A is the transcription/translation machinery, B is the DNA replication machinery, and C is the regulation (Fig. 4(b)).

Von Neumann's scheme implied a tape memory but as questioned by Elsasser,[6] what happens in cellular reproduction "it is a case of memory lacking the mechanistic side of storage. It is the ability to reproduce structural organization without invoking storage as a basic property." In other words, how can the complexity of a bacterium self-reproduce itself without apparent memory? The only memory is the DNA genome to produce proteins.

This nagging question was taken up by E. P. Wigner in 1962[7]: "the present writer has been baffled by the miracle that there are organisms which, if brought into contact with certain nutrient materials, produce

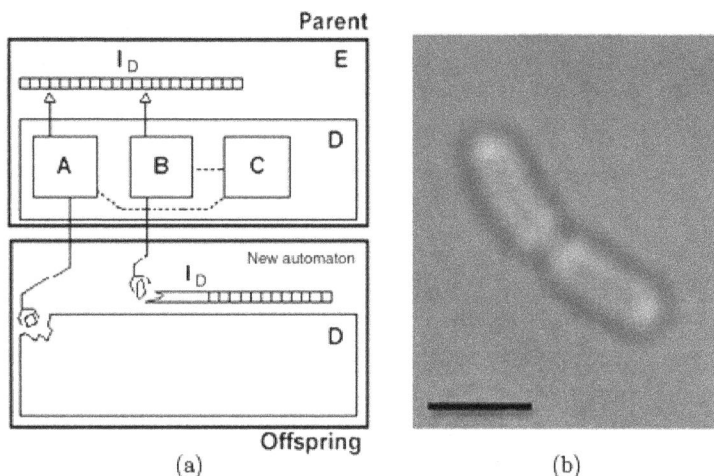

Fig. 4.   Von Neumann's logic for self-reproducing automaton. A is the Universal Constructor, B the copier, C the regulation, D the hardware, I_D the instruction. There is also an image of the division of an *E. coli* bacterium (scale $1\,\mu m$).

structures identical with themselves." He then showed that, according to standard quantum mechanical theory, the probability is zero for the existence of self-reproduction. Wigner's conclusion of his quantum computation is "defining by S the interaction between an organism V and a nutrient W, it is infinitely unlikely that there be any state of the nutrient which would permit the multiplication of any set of states which is much smaller than all the possible states of the system, unless S is tailored so as to permit reproduction." The real question is how to tailor S to permit reproduction.

Let us make some general remarks. In von Neumanns case, the formation of the offspring takes place outside of the parent body. Also the food given to the Universal Constructor undergoes no chemical change, otherwise analog elements would be introduced into a digital process.

In very simple terms, a self-reproducing machine is a finite array of components in a checkerboard so designed that it will grasp components from the conveyor belt and assemble them to form an exact replica of itself.[8]

Let us try now to motivate our experimental approach to self-reproduction. *In vitro* a cell originates from a cell,[9] so it is difficult to unravel this chicken and egg problem. So we decided with Vincent Noireaux to build an artificial cell and see how far we could go and how difficult it would be to let our artificial cell self-reproduce following a DNA program. In short, our philosophy was, "if you build it, you will know."

However, complex living cell organization and functions are, it is today conceivable to synthesize a cell from its basic elements. This approach aims

at a global understanding of cellular life, in particular the cooperative link
between its three essential components: the DNA information, the compartment, and the metabolism. The synthetic cell, built from scratch, would
be a unique compartment with a structure and an organization similar to
a bacterium. ATP and GTP would be used at the energy source in the
first stages of the development. For the information part, the synthetic
DNA programs would be expressed with the transcription and translation
machineries extracted from an organism. The physical boundary of the artificial cell would be a phospholipid bilayer. In aqueous solutions, phospho-
lipids self-assemble spontaneously into cell-sized vesicles, a process driven by
the hydrophobic interaction between the fatty linear chains. Lipid bilayers
are also the natural template for membrane proteins. Membrane proteins,
essential to any living system, carry out exchanges of materials and information between (Fig. 5).

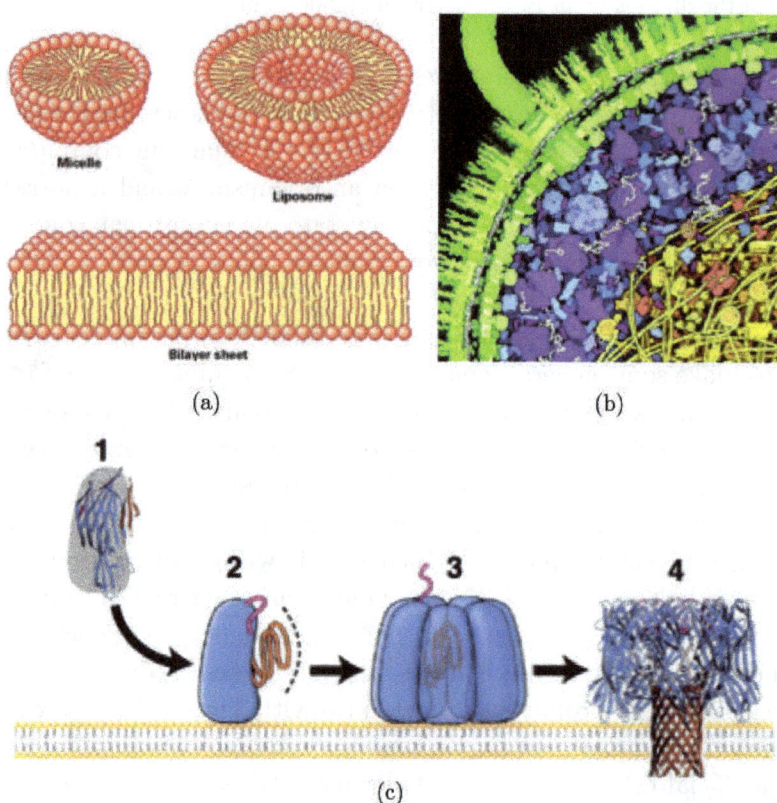

Fig. 5.   (a) Schematic of micelles and liposomes; (b) schematic of *E. coli* bacterium showing crowding; (c) self-organization of α-hemolysin proteins on a lipid membrane, forming
a pore.

Although useful to understand the global architecture of a self-reproducing cell, von Neumanns concepts do not capture the entire essence of cellular life. In bacteria, the DNA program represents only 1% of the volume of the organism. Life is not only a program; it relies also on many other fundamental non-genetic properties. Molecular self-organization and molecular crowding, for example, are critical processes in living cells. The phospholipid membrane displays a wealth of properties necessary for cellular organization, such as the formation of domain patterns. The phospholipid bilayer is, in turn, a template for the self-organization of macromolecular protein structures. In *Escherichia coli* , cross-linked networks and polymers responsible for cell rigidity and cell shape are assembled from the lipid bilayer. At the cellular level, one can wonder how a DNA program is developed to take advantage of self-organization processes. Assembling a synthetic cell unfolds the important of physical aspects that are, *in vivo*, regulated by already evolved gene networks. For instance, osmotic pressure favors exchanges of small nutrients by increasing the permeability of the lipid bilayers, the inside and outside world of the cell. Water is evidently an essential part of life, for vesicle formation as well as for molecular interactions and protein folding.

DNA is a chemically and mechanically robust biopolymer that serves as a code and a memory. DNA has almost no enzymatic or catalytic activity. In living cells, as well as in a DNA-programmed artificial cell, proteins perform the tasks, each one being encoded by a DNA gene. These nanomachines have a wide range of functions. Proteins can be catalysts, molecular motors, or membrane channels. Proteins self-assemble in large structures to carry out complex functions, such as flagella for motility. Proteins also interact physically with DNA to regulate gene expression. The transfer of the DNA sequence information to proteins is carried out in two steps: the transcription and the translation. During transcription, an RNA polymerase protein copies double stranded DNA into a single stranded messenger RNA. Transcription is rather simple compared to translation, which requires on the order of 100 proteins and tRNA molecules. The main component of translation is the ribosome, a large macromolecule that translates messenger RNA into proteins, using tRNA for the translation of one codon at a time. In bacterial cells, the couples transcription/translation process takes on the order of 1 min for a gene of 1,000 base pairs, from the beginning of transcription to the synthesis of a functional protein. DNA genes encode all of the transcription and translation components. Transcription and translation machineries, common to all living organisms, are the key part of a universal constructor, which can construct all of its own components. In an artificial cell, the

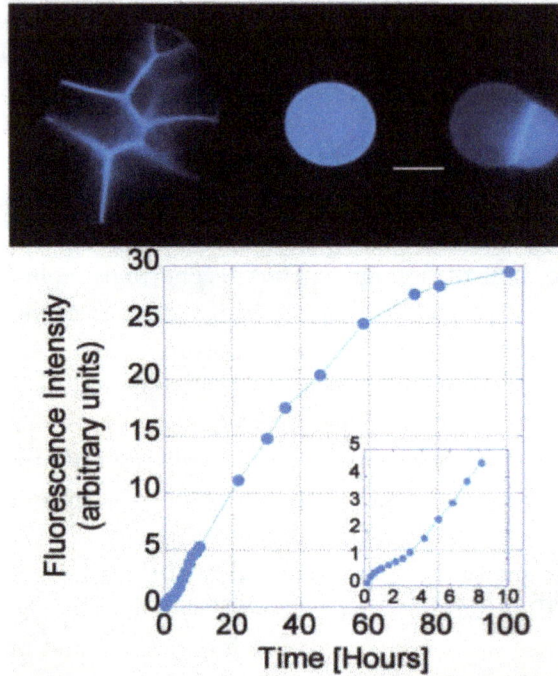

Fig. 6.   (a) Protein expression in a vesicle as a function of time (b) fluorescent images of α-hemolysin showing vesicles doublet, singlet, aggregate.

transcription/translation machineries must be physically present to carry out the expression of the first DNA programs.

Without entering into technical details, presented in Ref. 9, we were able to develop an artificial cell that could express proteins for about a week (Fig. 6). Now comes the difficult problem of self-reproduction. Two approaches are tested: the first one genetic, the second one physical.

In the first one, we use a purely genetic program following the bacterium *E. coli* as a model. Cytoskeleton proteins contribute to the mechanical stiffness of bacteria and are involved in essential cellular functions, such as cell division. One essential protein MreB is involved in cell shape and we showed that we could polymerize MreB on the wall of an artificial cell[10] (Fig. 7). Another protein FtsZ develops rings in *E. coli* which contraction leads to cell division. It has been implemented in artificial cells.[11] We will try to regulate the expression of those two essential genes to study how a vesicle can split into two.

In the second approach we follow a process of cell division occurring for spherical bacteria, called L-formed bacteria. There the cell produces

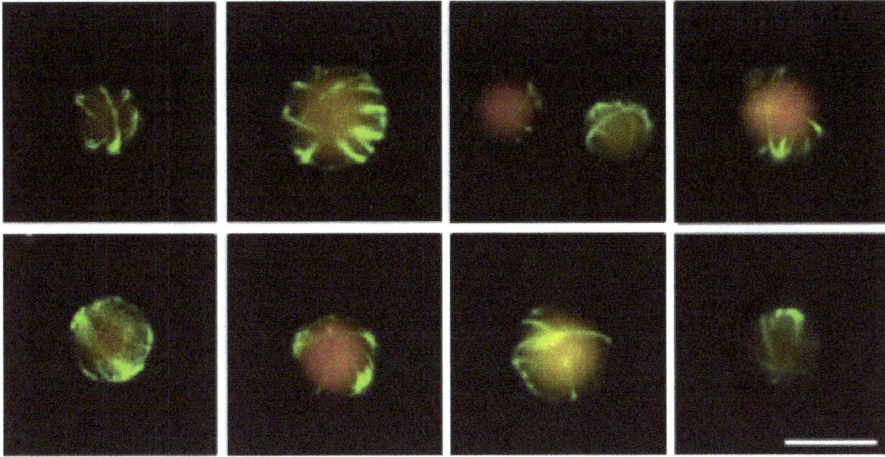

Fig. 7. Expression of MreB on liposomes. MreB proteins on the liposome surface form filamentous structures. It is an example of cytoskeleton protein polymerization. Scale bar 10 μm.

fatty acids that easily incorporate in the membrane of the cell. For a large production of fatty acids, the cell wall becomes unstable to the formation of many smaller cells, as shown by Errington.[12]

Fatty acids are easily incorporated in the membrane of vesicles, as long as the vesicles osmotic pressure is high enough (Fig. 8). We face the problem of 2-dimensional membrane growth that must be compensated by 3-dimensional vesicle volume growth for stability. Julien Dervaux in our laboratory has shown that for a high enough concentration of fatty acids injected from the outside world to the vesicle, the compensation cannot occur, water does not enter fast enough, the vesicle becomes unstable. It breaks into smaller vesicles (Fig. 8). In Figs. 8B–8D, one can observe three stages of destabilization ending with many connected vesicles.

Let me finally discuss recent studies by Pradeep Kumar in our laboratory[13] on the response of bacteria to high hydrostatic pressures (above 300 Atm). At these pressures proteins tend to destabilize. Amino acids lose some of their hydrophobicity. The first result is that protein polymerization is affected, the cytoskeleton destabilizes and the machinery of cell division breaks down. All the proteins are indeed produced, DNA replicates normally but cells cannot divide so dramatically elongated cells are produced (Fig. 9). When brought back to atmospheric pressure the elongated cell, after a cascade of divisions, gives cells back to their normal size. This experiment shows how delicate self-reproduction is in nature. The environment plays an essential role, for example, the local pressure. Let us note that such

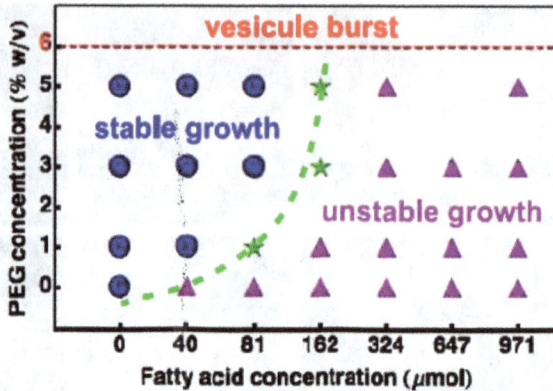

(A)

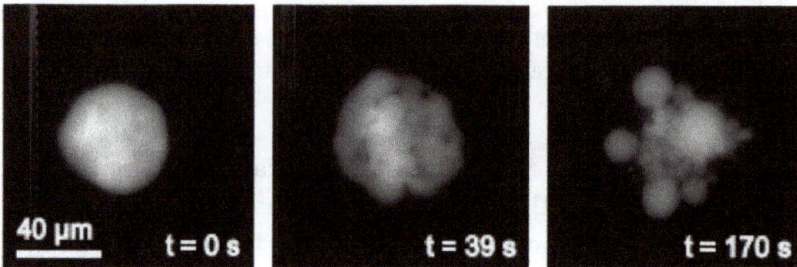

(B)   (C)   (D)

Fig. 8.   Phase diagram of vesicles in (A) Vesicle osmotic pressure (PEG concentration) versus fatty acid concentration. (B), (C) and (D) show the time evolution of vesicles destabilization when a large concentration of fatty acids is present.

pressures exist deep in the ocean where thermal vents are present with a huge ecosystem of bacteria around the vents.

In conclusion, we have constructed a minimal cell from scratch using the most powerful and versatile cell-free transcription/translation platform currently available. Encapsulated into phospholipid vesicles, the cell-free system is used to program cell-sized compartments with gene circuits towards division and self-replication. The theoretical difficulty lies in the coupling of digital and analog algorithms in biology and biochemistry. Partly digital at the chemical level (DNA, gene networks, distinct molecular configurations) and mainly analog in the dynamics of the self-reproduction process, biology implies a coupling between two far apart mathematical tools. Understanding the complex feedback loop implied by self-reproduction is a major undertaking.

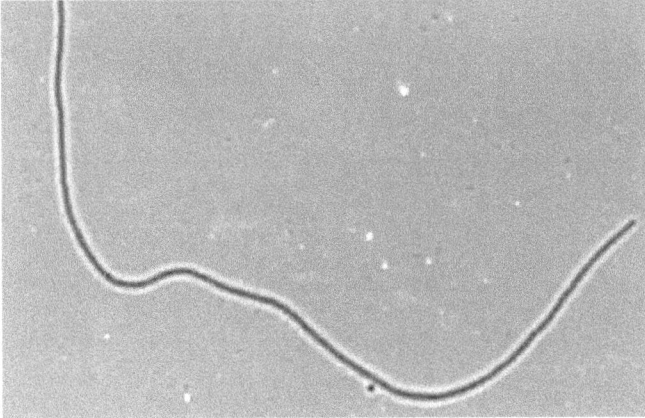

Fig. 9.   Elongated *E. coli* bacterium at a pressure of 300 Atm. The thickness of the filament is about 1 μm, a typical *E. coli* size.

References

1. E. Schrödinger, *What Is Life? The Physical Aspect of the Living Cell and Mind* (Cambridge University Press, 1943).
2. W. Harvey, *Exercitationes de generatione animalium* (Du-Gardianis, London, 1651).
3. R. Hooke, *Micrographia: Or Some Physiological Descriptions of Minute Bodies Made by Magnifying Glasses* (J. Martyn and J. Allestry, London, 1665).
4. R. Virchow, *Die Cellularpathologie: in ihrer Begründung auf physiologische und pathologische Gewebelehre* (Verlag von August Hirschwald, Berlin, 1858).
5. J. von Neumann, The general and logical theory of automata, *Cerebral mechanisms in behavior* **1**(1), 1–41 (1951).
6. W. Elsasser, *The Physical Foundations of Biology* (Pergamon Press, London, 1958).
7. E. Wigner, *The Probability of the Existence of a Self-reproducing Unit* (Routledge & Kegan Paul, London, 1961).
8. C. Langton, Self-reproduction in cellular automata, *Physica D* **10**(1), 135–144 (1984).
9. V. Noireaux, Y. Maeda and A. Libchaber, Development of an artificial cell, from self-organization to computation and self-reproduction, *Proceedings of the National Academy of Sciences* **108**(9), 3473–3480 (2011).
10. Y. Maeda *et al.*, Assembly of MreB filaments on liposome membranes: a synthetic biology approach, *CS Synthetic Biology* **1**(2), 53–59 (2011).
11. M. Osawa and H. Erickson, Liposome division by a simple bacterial division machinery, *Proceedings of the National Academy of Sciences* **110**(27), 11000–11004 (2013).
12. R. Mercier, Y. Kawai and J. Errington, Excess membrane synthesis drives a primitive mode of cell proliferation, *Cell* **152**(5), 997–1007 (2013).
13. P. Kumar and A. Libchaber, Pressure and temperature dependence of growth and morphology of *Escherichia coli*: experiments and stochastic model, *Biophys. J.* **105**(3), 783–793 (2013).

Spin Glasses and Frustration

Scott Kirkpatrick

School of Engineering and Computer Science
Hebrew University
91904 Jerusalem, Israel
kirk@cs.huji.ac.il

The 1975 EA paper by Sam Edwards and Phil Anderson proposed to understand a class of real materials as frustrated, glassy magnets with a novel type of weak order in time. It launched an exciting period of intense exploration of new directions for understanding random and composite materials. This work has had a lasting impact well beyond materials science, contributing powerful new methods for optimization of complex many-parameter systems as well as codes offering Shannon-bounded communications in extremely noisy channels.

1. Origins of the EA (and SK) Papers

This paper will start with a bit of personal memoir of some exciting times in condensed matter physics, the late 1960s and the 1970s, in which the paradigm for describing real materials — alloys, compounds, composites and glasses — shifted from a search for the idealized uniform "coherent potential" or "effective medium" to an effort to tease out the novel consequences of quenched disorder, such as percolation, localization, and glass transitions into amorphous but highly stable structures. These regimes, where 30+ years of conventional theory had broken down, seemed to be where Phil Anderson felt most alive, so of course he played a highly visible and central role in all of this. Other speakers cover percolation and localization; I get to talk about spin glasses. In these first sections I will stick to the period 1974 to 1978 and a bit later, in which the basic picture evolved rapidly. A final section covers the many ways in which ideas and technical methods that arose in this period of ferment are being employed today.

Alloy a few Mn atoms in a matrix of pure Cu (or almost any transition metal in a noble metal host) and you get long-ranged magnetic interactions, oscillating in sign, a Kondo effect, and a puzzling set of possibilities for the magnetic structures that might be found at higher densities. Increasing the concentration of the transition metal component to a few per cent, as crystal growers at Bell Labs, University College London, Grenoble's CRTBT and others soon did, yielded anomalous thermal and magnetic susceptibilities with cusps rather than sharp peaks, and magnetic hysteresis effects on very long time scales. For the experimental facts and debates, see Joffrin[1] for a view at the time, and Mydosh[2] for a view from a few years removed. Either Phil or Brian Coles (accounts differ on this) suggested the name "spin glass," to capture the slow hysteretic effects of these materials and others with mixtures of ferromagnetic and antiferromagnetic interactions. That name had a strong appeal.

The Edwards–Anderson (EA) paper[3] captured the main ideas which underscored that appeal. They made explicit their goal of studying the spin glass in the hopes of learning more about the transition in structural glass. First, they introduced an order parameter which measured stability as the persistence of the value of a spin, without specifying the details of the spins' arrangements:

$$q = (1/N) \sum_i \langle \langle S_i \rangle_T \langle S_i \rangle_T \rangle_{disorder}.$$

Note the separation of the thermal average (the long term stability in the presence of the sources of randomness, quenched in fixed positions) from the average $\langle \ldots \rangle_{disorder}$ over the possible positions of the magnetic atoms. The EA paper then brought together a set of technical tricks that could lead to a mean field theory for the onset of such order. The first was a replica construction in which the spin average above is performed by comparing two or more identical but non-interacting copies of the disordered material under study. Second was the trick of evaluating thermal averages of the free energy of such a quenched random system by taking the limit as $n \to 0$:

$$\langle \langle log\ Z \rangle_T \rangle_{disorder} = \langle lim_{n \to 0} (Z^n - 1)_T / n \rangle_{disorder}.$$

The results made it seem obvious that, at least in mean field theory, such a model could exhibit a phase transition into a new spin glass phase.

Phil later described[4] these manipulations as "a mathematical trick so hoary that its origins are lost in history." They can be found, for example, in a 1959 paper on dilute ferromagnets by Robert Brout[5] which appears to have been written while he was spending a summer at Bell Labs. (Thanks to Giorgio Parisi for the reference.)

I read the EA paper as soon as David Sherrington showed up at IBM Watson Labs for the summer of 1975. He had been in contact with Sam Edwards during the EA paper's gestation, and had an idea. We could keep the general approach of EA, but change the model to make the whole thing "solvable," and thus end up with a rigorous mean field limiting version of the actual problem. Mn and Fe ions in noble metals look like idealized Heisenberg spins, but Ising spins are easier to deal with. RKKY interactions are complicated, so like EA, we just replaced them with a Gaussian distribution of exchange constants, but let the range become infinite — spherical models, with N spins each interacting with the other $N - 1$ spins are solvable when ferromagnetic, right? And the classic spin glasses become ferromagnets when the transition metal concentration exceeds about 10%, so we added a term favoring ferromagnetic alignment, and ended up with the generalized Hamiltonian

$$H = \sum_{\text{all } i,j} J_{ij}(S_i S_j)/\sqrt{N} + J_0 \sum_i S_i/N$$

where the J_{ij}, from a Gaussian distribution with unit variance, are scaled as $1/(\sqrt{N})$ and J_0 by $1/N$, to keep things finite and in balance. After deriving a solution along the same lines as in EA enough times that we thought we had it right, we were left with a nonlinear equation for the free energy which could be solved at its critical points, around its ground state, and numerically in between. We made the classical assumption that evaluation of the partition function for n replicas of our model would be dominated by a single stable minimum or maximum about which a stationary phase expansion would capture the important contributions. This gave a plausible phase diagram, seen in Fig. 1, a susceptibility and specific heat with cusps, as seen in Fig. 2, and resembled much of the experimental data.[6]

There was, of course, one catch. The solution gave values for the ground state energy

$$E(T = 0)/N = -1/(2\pi) \approx -0.79$$

and the entropy

$$S(T = 0)/N = -\sqrt{2/\pi} \approx -0.17.$$

The first was plausible (it took two years of work on computer simulations at the speeds then available to show that it is too low by about 0.03), but the second result was unacceptable for an Ising model, whose discrete excitations should be frozen out at zero temperature. So we inserted warning labels on this part of the results, wrote a very understated abstract and sent the paper off. The physics seemed right, and the evidence for a phase transition out of the high-temperature paramagnetic phase into, well, something pretty interesting, was solid.

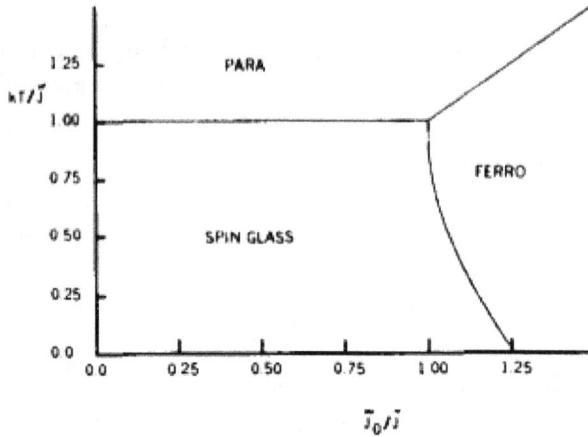

Fig. 1.   Phase diagram of a "solvable" spin glass.

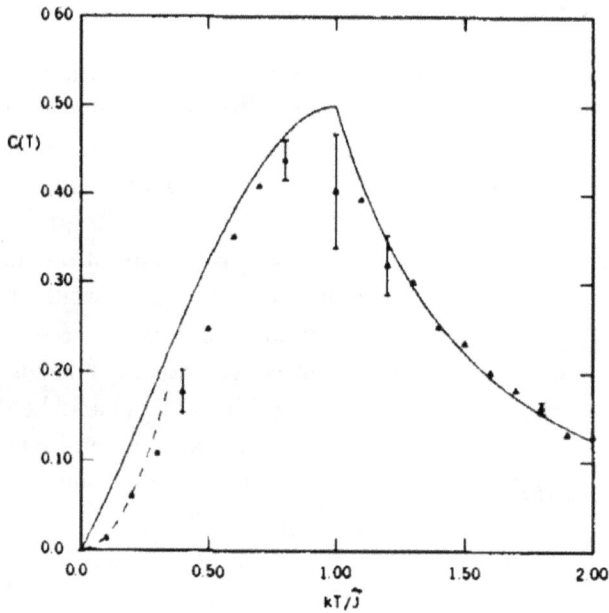

Fig. 2.   Specific heat of a "solvable" spin glass (with computed values from samples with 500 spins).

2. Cavities, TAP Equations, and All That...

The two papers were well received. Experimentalists whose data looked like our curves were happy, and theorists had a fat target to shoot at. A workshop was quickly arranged to be at Aspen for summer 1976, with PWA, David

Thouless and Richard Palmer at the center of things. The main objective was to find methods of solving this class of problems in mean field theory without the introduction of replicas and the very suspicious $n \to 0$ limit. Attention focused on developing "cavity methods" in which the force on the spin at a site is calculated by summing the effects of all its neighbors in the "cavity" left by removing that spin. The spin's response is then fed back into the problem for subsequent iterations. This idea for extending mean field theory goes back to Bethe, is now called "message passing" by statistical physicists, and is also known as "belief propagation" in the statistics literature.

My recollection is that in discussions over beer in Aspen the problem was solved multiple times on multiple evenings. Each evening resulted in an "Aspen preprint." In the light of the next day, each such preprint was found to have problems, calling for another day of discussion, another preprint. It was an exciting time. Phil was wont to mutter with satisfaction, as each preprint improved over its predecessor, "And gentlemen at Harvard, now abed, shall think themselves acurs'd they were not here." The results appeared in the Oct. 1976 TAP paper,[7] and gave different low temperature behavior, in rough agreement with the computer simulations on the model which were then becoming available.[8] See the dashed line for the low temperature specific heat extrapolation in Fig. 2.

Shortly after the Aspen workshop ended, Thouless and de Almeida found the fatal flaw in the replica calculation. It wasn't the $n \to 0$ limit, nor the several exchanges of the order of evaluation, but simply that the n-replica calculation of the partition function was not to be calculated around a single fixed point, since that point was neither a minimum nor a maximum. Replica symmetry had broken.[9] It took about six more years and a series of attempts of increasing complexity to develop a convincing hierarchical model of replica symmetry breaking for the SK model's low temperature phase[10] with interesting consequences and twenty-plus years beyond that to obtain a rigorous justification of the main features of that proposal.

But the idea of gathering researchers and students, experimenters and theorists, interested in glasses, random magnets, percolation, localization and conduction into one community had taken hold. Phil headlined an influential European summer school that lasted eight weeks at Les Houches in 1978, taking a break from the stress of receiving a major award over the previous winter. He covered all these subjects from the perspectives of single-particle spectra and many-particle effects in a magnificent series of lectures.[4] Because of its breadth of coverage and extreme duration, the summer school came to be known as "Camp Houches," a basic training experience for a whole generation of European (and some American) scientists still active in the study of complex and disordered systems.

3. The Spin Glasses of Today...

At this point the further evolution of the spin glass project begins to split into several distinct routes. The high road leads upwards to the various possible forms of replica symmetry breaking and their consequences. Less rarified paths lead through problems in combinatorics that also have natural large scale limits while representing problems with real world relevance. And orthogonal directions have now used message passing techniques stimulated by the TAP equations to create powerful codes, a large subject that I won't attempt to discuss here. (But you can learn about them, described from this perspective, in the recent book by Mezard and Montanari.[11])

In a series of papers by Parisi, various collaborators along the Paris-Rome axis, and others, increasingly elaborate schemes for breaking replica symmetry and evaluating $\langle\langle \log Z \rangle\rangle$ were explored, starting with one-step symmetry breaking in which the replicas were divided into m groups, with one order parameter found between different groups and another within a group. This was not good enough for the SK model, although it seems to work for certain problems in combinatorics and constraint satisfaction. The

Fig. 3.   Resting at Camp Houches, 1978.

ultimate form of replica order parameter for the SK model is a continuous function $q(x)$ on the line $0 \leq x \leq 1$, representing a continuous distribution of quasistable states of low energy. A surprising consequence emerged from this full RSB approach. Correlations between these "pure" states are ultrametric. If we calculate the pairwise overlaps between three states, A, B and C, e.g.:

$$Q^{A,B} = (1/N) \sum_i S_i^A S_i^B$$

then at least two of the overlaps, $Q^{A,B}$, $Q^{A,C}$, and $Q^{B,C}$, are equal. This is most simply understood as suggesting a clustered, tree-like structure in which sets of "pure" states are derived from one another. Since in a glass or a spin glass, we generally can observe only one state at a time, the experimental consequences of ultrametricity are unclear. But in large scale problems of searching or optimizing networked objects, multiple solutions can be obtained, so the predictions of clustering or ultrametricity prove important for establishing heuristics or characterizing the space of possible solutions and algorithm performance in that space.

The cavity approach, and the directed message passing suggested by the TAP equations, has had perhaps its widest application in the optimization and searching of complex systems, such as information networks. One simple problem has proven to be replica symmetric. Consider the simplest possible metric-free travelling salesman problem (TSP). Let the upper triangle of the $N \times N$ matrix of distances $d_{i,j}$ between each pair of N points be filled by random numbers from some distribution. The uniform distribution on the interval $[0, 1]$ proves most convenient. We symmetrize all distances so that $d_{i,j} = d_{j,i}$ and then ask for the shortest continuous tour of links connecting all sites without visiting any site more than once. A greedy solution, at each step choosing the link to the closest remaining unvisited site, leads to a tour length that diverges as $\log N$. Both a replica-symmetric Ansatz[10] and a cavity approach,[12] however, predict a constant limiting tour length as $N \to \infty$. The cavity solution is more easily evaluated, and predicts that the optimal tour will have length 2.0415... in the large N limit, in agreement with numerical experiments.

The environment in which message-passing methods have most obviously burgeoned is search on the Internet. Its scale today is set by the earth's population, a few billion. In the near future, an Internet filled by information from the world's sensors may increase in scale by many more decades. Finding the answers most relevant to a particular query in this vast space requires pre-conditioning each search with information passed selectively along the same network that is being searched. Today's search engines rely upon several families of methods for accumulating these hints, whether by identifying "hubs" (sources of many useful links) and "authorities" (generators of links most likely to be relevant to the specific query), or by ranking pages by the degree to which other pages link to them. For a good overall review of these approaches see Easley and Kleinberg's recent book.[13]

Searches for groups of similar objects or for communities with similar interests within this large space are important in building systems that make reasonable reccomendations or that establish trust. They may need to operate on scales of order N (the world's or a nation's population), $\sqrt{N}$ (a few hundred thousands) or even smaller. A classic problem to use in idealizing such a search is the problem of finding the largest clique (completely connected subgraph) in a simply defined random graph, such as the Erdös–Rényi $G(N, p)$ with N nodes and a fraction, p, of its bonds present. The realization that for, say, $G(N, 1/2)$ as $N \to \infty$, the largest clique that will be found has exactly $2 \log_2 N$ nodes was an early example of the recognition that sharp phase transition-like behavior occurs in large mathematical objects as well as in materials.[14] Finding even one such clique (there will be many) is extremely hard, since a greedy search will run out of nodes to search among once the clique found is of order $\log_2 N$, and no stronger search methods are known. So in this problem, an SK-like model has been identified and extensively studied. One simply "plants" a clique of size $\sqrt{N}$ in the graph and asks for an algorithm that can find it, preferably in time proportional to the number of links in the graph. At the moment, the best algorithm for this search uses a very simple form of message-passing. Deshpande and Montanari[15] use the power method, raising the graph's adjacency matrix to a small power, to find its largest eigenvalue and the associated eigenvector. The large components of that eigenvector are the elements in the planted clique. I am currently seeing if this trick of using messages to amplify the weak signal that a clique gives off can be pushed down to the $\log_2 N$ scale where naturaly occurring cliques are found, by combining the power method trick with a steady reduction in the scale of focus.

Finally, an important indirect consequence of the spin glass project (to me, at least) is the realization of simulated annealing as an archetype of a wide range of physically-motivated methods of searching for optimal solutions to complex non-convex problems with many parameters. While spending a year or so searching among the SK model's low energy configurations to determine ground state energies and spin relaxation dynamics, it became perfectly natural for me to let samples evolve at finite temperatures using the Metropolis procedure for the dynamics. Heating the sample up and slowly cooling it was a natural way to improve on ground state energy estimates and to obtain good statistics. In subsequent years, it was equally natural for a small group of us at IBM interested in designing good automatic tools for placing and routing computer components during the design process to view our problems as medium sized spin glasses. There was frustration due to the conflict between the need to optimize performance by putting circuits as close to one another as possible, and the constraints of minimizing heat generation and making wiring possible, both of which tended to force circuits apart. When we sent off a paper summarizing half a dozen such examples and stating this approach as generally as possible, we encountered a certain amount of resistance to our unconventional subject and intuitive approach. Phil, however, fully appreciated the ideas, liked the paper, and was instrumental in getting it accepted in *Science*.[16]

Acknowledgments

The spin glass community is very much alive today, well into its third generation. While preparing this article I had the pleasure of taking part in 2014's "Spin Glasses: an old tool for new problems," at the Institut d'Études Scientifique de Cargèse, organized by Lenka Zdeborova, Federico Ricci-Tersenghi, Florent Krzakala, and Giorgio Parisi. Work reported there ran the gamut from studies of glasses (jamming and dynamics in hard sphere models) to information theory in engineering, computation and biology to rigorous proofs, some recently obtained, of replica solutions to the originally-stated problems.

References

1. J. Joffrin, Disordered systems — Experimental viewpoint, in R. Balian, R. Maynard and G. Toulouse (Eds.), *Les Houches 1978: Ill-Condensed Matter* (North-Holland, Amsterdam, 1978), pp. 63–157.
2. J.A. Mydosh, *Spin Glasses: An Experimental Introduction* (Taylor and Francis, London, 1993).
3. S.F. Edwards and P.W. Anderson, *J. Phys. F* **5**, 965 (1975).
4. P.W. Anderson, Lectures on amorphous systems, in R. Balian, R. Maynard and G. Toulouse (Eds.), *Les Houches 1978: Ill-Condensed Matter* (North-Holland, 1978), pp. 159–261.
5. R. Brout, *Phys. Rev.* **115**, 824–835 (1959).
6. D. Sherrington and S. Kirkpatrick, *Phys. Rev. Lett.* **35**, 1792 (1975).

7. D.J. Thouless, P.W. Anderson and R.J. Palmer, *Philos. Mag.* **35**, 593–601 (1976).
8. S. Kirkpatrick and D. Sherrington, *Phys. Rev.* **17**, 4384–4403 (1977).
9. J.R.L. de Almeida and D.J. Thouless, *J. Phys. A* **11**, 983–990 (1978).
10. M. Mezard, G. Parisi and M.A. Virasoro, *Spin Glass Theory and Beyond* (World Scientific, 1987).
11. M. Mezard and A. Montanari, *Information, Physics, and Computation* (Oxford University Press, 2009).
12. W. Krauth and M. Mezard, *Europhys. Lett.* **8**, 213–218 (1989).
13. D. Easley and J. Kleinberg, *Networks, Crowds and Markets: Reasoning about a Highly Connected World* (Cambridge University Press, 2010).
14. D.W. Matula, *Proc. Chapel Hill Conference on Combinatorial Math. and Its Applications*, Univ. North Carolina, 1970, pp. 356–369.
15. Y. Deshpande and A. Montanari, arXiv:1304.7047 (2013).
16. S. Kirkpatrick, C.D. Gelatt, Jr. and M.P. Vecchi, *Science* **220**, 671–680 (1983).

Frustration and Fluctuations in Systems with Quenched Disorder

D. L. Stein

Department of Physics and
Courant Institute of Mathematical Sciences
New York University
New York, NY 10003, USA
daniel.stein@nyu.edu

As Phil Anderson noted long ago, frustration can be generally defined by measuring the fluctuations in the coupling energy across a plane boundary between two large blocks of material. Since that time, a number of groups have studied the free energy fluctuations between (putative) distinct spin glass thermodynamic states. While upper bounds on such fluctuations have been obtained, useful lower bounds have been more difficult to derive. I present a history of these efforts, and briefly discuss recent work showing that free energy fluctuations between certain classes of distinct thermodynamic states (if they exist) scale as the square root of the volume. The perspective offered here is that the power and generality of the Anderson conception of frustration suggests a potential approach toward resolving some longstanding and central issues in spin glass physics.

1. Phil Anderson and Spin Glass Theory

It is a great pleasure, both personally and scientifically, to contribute to this volume in honor of Phil Anderson's 90th birthday. The importance and influence of Phil's research in shaping the modern field of condensed matter physics (including coining the term, along with Volker Heine) is widely recognized. There are few currently active, fruitful areas of condensed matter research that have not been either created or (at least) strongly influenced by Phil. His influence, moreover, is not limited to condensed matter physics: he pointed the way[1] toward what is now universally known as the Higgs mechanism; set the stage for later developments in complexity science by emphasizing the importance of a non-reductionist scientific viewpoint[2] (not a widely held view at the time); and later explored and emphasized the connections between the statistical mechanics of quenched disorder and problems in biology, graph theory, and other areas outside of physics.[3]

It is the topic of quenched disorder that I will address here. I will not discuss its applications to other areas, but will instead return to the theory's roots. The subject of disordered systems — in particular glasses and spin glasses — has been a longstanding interest of Phil's, and one in which he has made numerous fundamental contributions. At the height of his interest, he identified the problem of understanding the physics of structural glasses and their magnetic counterparts, spin glasses, as one of the central unsolved problems in condensed matter physics.[4] His views may well have changed in the ensuing decades, but one can safely argue that understanding the effect of quenched randomness on the condensed state has presented one of the most persistent and confounding set of problems in modern condensed matter physics.

On the subject of spin glasses proper, Phil's contributions are too numerous to list, but it would be a dereliction of duty not to mention two of his most foundational papers. The first is well-known: his 1975 paper[5] with Sam Edwards that swept away numerous distracting details and identified quenched, conflicting ferromagnetic and antiferromagnetic interactions as the ultimate microscopic basis of spin glass behavior. Edwards and Anderson (hereafter referred to as EA) used this idea to propose a simplified model Hamiltonian that has since formed the basis of most theoretical investigations (we include here studies of the Sherrington–Kirkpatrick model,[6] which is an infinite-range version of the EA model). The other important idea proposed in the EA paper concerned the nature of the spin glass order parameter, but that's less relevant to the discussion below.

The second of these papers, not as widely known or cited, concerns Phil's joint (with Gérard Toulouse) introduction of the concept of frustration. The story goes that Gérard attended a lecture in 1976 in which Phil wrote on the blackboard, "The name of the game is frustration." Whether this was elaborated on in the talk I couldn't say, and the principals will have to provide the details — if they remember. But it would have been characteristic of Phil to make this cryptic remark without elaboration and move on. Inspired, Gérard published a classic paper the next year[7] that remains the canonical definition — both conceptual and operational — of frustration. In this formulation, one considers the spins at lattice sites that form a closed loop on a lattice. If there are an odd number of antiferromagnetic couplings on the edges constituting the loop, the spins cannot be arranged to satisfy all of the interactions.

The following year, Phil published an alternative, and more general, definition of frustration[8] in which one studies the free energy fluctuations of two blocks of material (glass, spin glass, ferromagnet, what have you) that have independently relaxed to their respective ground states. I will elaborate both on this and the Toulouse definition of frustration in Sec. 3. For now I will just note that this latter approach has not received as much attention as Toulouse's, but nonetheless — as usual for Phil — it is enormously prescient. In fact, I will argue below that this alternative approach to frustration may, after a long period of dormancy, contain just the right perspective to resolve some longstanding open questions at the heart of spin glass physics.

But for now, the main point is that in both cases frustration arises when a system contains many fixed, conflicting internal constraints, not all of which can be simultaneously satisfied. Like art, physics can sometimes imitate life.

The final topic of this informal account concerns Phil's role in the naming of spin glasses. Phil didn't name them himself, although he clearly was involved.[9] Regardless of the details, he was as usual present at the creation, and provides an amusing discursion on the early days. Most accounts, including Phil's, credit Bryan Coles with inventing the term "spin glasses", although details among the accounts vary somewhat. A competing (and in my opinion, more ungainly) term that had gained some traction at the time was "mictomagnetism", and we might well today be referring to mictomagnets rather than spin glasses.[a] Phil's dry sense of humor led to a disquisition on the etymology of the term, and one cannot do justice to his description other than quoting it in full[9]:

> "A few weeks ago I received a letter from Ralph Hudson of the NBS objecting to this term, on the basis that he thought that the only other word in the English language using the same root was 'micturation' and the root was Latin for 'urine'. I think myself that the term is very descriptive: back in the Middle West we used to refer to something as 'p—poor' if it was not worth anything more substantial, and that is a good description of this kind of magnetism. The remanence is small and sluggish, there are peculiar training phenomena, and the susceptibility is often very history dependent. Unfortunately, I am assured by Collin Hurd and by the OED that Hudson is incorrect and that 'micto' is a legitimate Greek root meaning 'mixed'. "

---

[a]In fact, the term still survives, and you can google it, although Google will insist that you must have meant "micromagnets." Don't give in.

Those interested in Phil's first-hand perspective of the early days of spin glass research should consult his series of *Physics Today* "reference frame" articles[10–16] that helped bring the subject to the attention of the broader physics community.

2. Free Energy Fluctuations in the Random Field Ising Model

Systems with quenched disorder possess several features distinguishing them from homogeneous systems. Our focus is on one of these: their energies and free energies are random variables depending on the disorder, and their concomitant fluctuations contain information that can potentially resolve central open questions that remain intractable to this day. These questions include the conjectured multiplicity of pure and ground thermodynamic states, the relations between such distinct states (if they exist), the geometry and energy scaling of their relative interfaces, and so on.

To illustrate the potential usefulness of the information provided by these fluctuations, we turn briefly to a different system: the random-field Ising model (RFIM), which is a uniform Ising ferromagnet subject to a random external field. It can be modelled using the Hamiltonian

$$\mathcal{H}_h = -\sum_{\langle x,y \rangle} \sigma_x \sigma_y - \epsilon \sum_x h_x \sigma_x. \tag{1}$$

Here x and y are lattice sites in the d-dimensional cubic lattice $\mathbf{Z}^d$, $\sigma_x = \pm 1$ is an Ising spin at site x, the first sum is over nearest-neighbor pairs of sites only, and the fields h_x are independent, identically distributed random variables representing local external fields acting independently at each site x. For simplicity, we take the probability distribution of the h_x's to be Gaussian with mean zero and variance one. The subscript h on the LHS of Eq. (1) refers to a particular realization of the h_x's.

One can now ask whether the ferromagnetic ground state is unstable to breakup by the random field. The answer is clearly yes if ϵ is sufficiently large. But is it true for *any* nonzero ϵ?

For a uniform field this question is trivial: in any dimension, a field of any fixed, nonzero magnitude determines the magnetization direction at all temperatures, and so there is no phase transition. A simple scaling argument explains why. In zero field below T_c, consider the positively magnetized (i.e., "up") phase. Now apply a small uniform field of magnitude h pointing down. Overturning a compact patch (or "droplet") of spins of length scale L to align with the field is energetically favorable for sufficiently large L: for

Ising spins, the cost in surface energy is of order L^{d-1} while the lowering of bulk energy is of order hL^d. So, no matter how small h is, the system can lower its energy by overturning a sufficiently large droplet. At positive temperature, overturning droplets of spins is certainly entropically favorable as well. Consequently, in any dimension and at any temperature (including zero), there is a unique Gibbs state in an external uniform field of any nonzero magnitude.

Now consider the case when quenched disorder is present, i.e., when the field is random. This requires a modification of the above argument, which was provided in 1975 by Imry and Ma.[17] The boundary energy, which depends only on the ferromagnetic couplings, is unchanged. The bulk energy, however, is determined by the fields, which now fluctuate from region to region. Nevertheless, in an arbitrary large droplet containing L^d spins, the central limit theorem requires that the typical bulk energy scales as $L^{d/2}$. So in two dimensions the competing boundary and bulk energies scale similarly with volume. Imry and Ma concluded that in two dimensions and below, the ferromagnetic ground state should be unstable for any nonzero ϵ, while above two dimensions (and with small but nonzero ϵ) ferromagnetic long-range order persists. (In fact, Imry and Ma mainly focused on continuous spin models, where the boundary energy scales as L^{d-2}, giving a lower critical dimension of 4.)

Normally, such an argument would be sufficient to settle the matter, but a few years later a more detailed field-theoretical analysis based on supersymmetry[18] concluded that the critical behavior of a d-dimensional spin system in a random external field is equivalent to that of the corresponding $(d-2)$-dimensional system in the *absence* of an external field. This "dimensional reduction" argument therefore predicts the lower critical dimension of the RFIM to be three.

This controversy was eventually resolved by rigorous mathematical arguments, first by Imbrie[19] and later by Aizenman and Wehr.[20,21] Imbrie proved that the Ising model in a random magnetic field in three dimensions exhibits long-range order at zero temperature and sufficiently small disorder, indicating that the lower critical dimension of the RFIM is indeed two. Aizenman and Wehr later proved that in two dimensions at all temperatures and fields, the RFIM possesses a unique Gibbs state.

This now ancient controversy is recounted for two reasons. First, it represents an interesting — and rare — example where rigorous mathematics resolved an open and important controversy in theoretical (and indeed, experimental) physics. It may be the case that something similar will be required for spin glasses and possibly even structural glasses, where the most basic questions have persisted as a subject of intense controversy over decades.

More relevant to this paper, though, is that Aizenman and Wehr essentially made the Imry–Ma argument rigorous by analyzing fluctuations (with respect to the quenched disorder) of the free energy difference between putative positively and negatively magnetized states. Although the RFIM is not a frustrated system, the spirit of the Aizenman–Wehr method aligns with the Anderson approach to characterizing and understanding frustration.

3. Frustration

We turn now to finite-dimensional spin glasses, where almost all of the basic questions remain open. These include whether an equilibrium phase transition occurs above some dimension; if so, the nature of the broken symmetry (if any) of the spin glass phase; whether (up to global symmetry transformations) the spin glass phase is unique; if not, the nature of the relationships among the many spin glass phases; whether there exists an upper critical dimension above which mean field theory holds;[b] and numerous others. And this list doesn't include questions concerning the non-equilibrium dynamical behavior of spin glasses, which won't be addressed in this paper.

For concreteness we confine our attention to nearest-neighbor models defined by the EA Hamiltonian:[5]

$$\mathcal{H}_{\mathcal{J}} = -\sum_{\langle x,y \rangle} J_{xy}\sigma_x\sigma_y - h\sum_x \sigma_x \,, \tag{2}$$

where x and y are sites in the d-dimensional cubic lattice, $\sigma_x = \pm 1$ is the Ising spin at site x, the couplings J_{xy} are independent, identically distributed random variables, $\mathcal{J}$ denotes a particular realization of the couplings (corresponding physically to a specific spin glass sample with quenched disorder), h is an external magnetic field, and the first sum is over nearest-neighbor sites only. We hereafter take $h = 0$ and the spin couplings J_{xy} to be symmetrically

---

[b]In homogeneous systems, this question usually refers only to behavior at or near the critical point. For spin glasses it is considerably more far-reaching. Here we're asking whether the *low*-temperature properties — i.e., the order parameter and the nature of broken symmetry — corresponds in *any* finite dimension to the replica symmetry breaking[22–26] that occurs in the low-temperature phase of the infinite-range Sherrington–Kirkpatrick model.[6] This is generally not an issue in homogeneous systems, where mean-field theory usually provides a useful guide to the nature of the low-temperature phase well below T_c in any dimension where a phase transition occurs.

distributed about zero; consequently, the EA Hamiltonian in (2) possesses global spin inversion symmetry.

A striking feature of the EA Hamiltonian is the presence of *frustration*, meaning the inability of any spin configuration to simultaneously satisfy all couplings. It is easily verified that, in any dimension larger than one, all of the spins along any closed circuit C in the edge lattice cannot be simultaneously satisfied if

$$\prod_{\langle x,y\rangle\in C} J_{xy} < 0. \tag{3}$$

This definition of frustration is due to Toulouse.[7]

Toulouse's geometry-based definition is appealing on several levels, and has been the starting point for numerous investigations (see, for example, Refs. 27 and 28). It provides a simple test to determine whether a given type of spin system possesses frustration, and suggests the underlying reason why certain systems may possess multiple pure or ground states. More generally, the quantification of conflicting internal constraints provides a powerful conceptual tool for understanding certain general aspects of complex behavior in the broader study of complex systems.

Its drawback is that in some sense the Toulouse approach is *too* well-suited to spin glasses; it is difficult to see how it can be generalized in a natural way to non-spin systems that surely possess frustration, such as structural glasses or combinatorial optimization problems. For these systems the Anderson definition of frustration is more useful; it is sufficiently general that (with minor modification as needed) it should apply to any system. As we will see, it also provides a conceptual starting point for mathematical studies that hold promise for resolving the long-controversial issue of pure/ground state multiplicity in frustrated systems with quenched disorder.

The idea itself is rather simple, although its simplicity conceals a profound and very useful insight. Based on a preliminary study of Anderson and Pond,[29] Anderson proposed[8] considering the free energy fluctuations of two statistically identical blocks of the same material. It is simplest to describe the procedure at zero temperature, although it is easily modified for positive temperature. So let each block of material independently relax to its ground state.[c] One then brings the two blocks together and measures the fluctuations in the coupling energy across their interface. In non-frustrated systems, such as ferromagnets, the energy fluctuations scale as the surface area A of

---

[c]For a finite system with specified boundary conditions, such as periodic, and continuous disorder, such as Gaussian, the ground state is unique up to a global symmetry.

contact. This scaling holds for both homogeneous and random ferromagnets (in which the bond strengths are positive, i.i.d. random variables).

However, if frustration is present, then it will be the case that

$$\lim_{A\to\infty} \langle E^2 \rangle / A^2 = 0. \tag{4}$$

In fact, one can turn this around and use (4) as the general definition of frustration, which we will hereafter do.

Using reasoning based on the central limit theorem Anderson further conjectured that

$$\lim_{A\to\infty} \langle E^2 \rangle / A = O(1). \tag{5}$$

That is, in a frustrated system one might expect the energy fluctuations of the ground states to scale as the *square root* of the surface area of contact. However, (5) is a rough estimate.

The definition (4) of frustration, although not as widely known or appreciated as (3), contains the seeds of a powerful approach to understanding realistic spin glasses and other complex systems. We turn now to a natural outgrowth of this approach, namely, scaling theories of the spin glass phase.

4. Scaling Theories of the Spin Glass Phase

The idea of investigating fluctuations of free energy differences in the presence of frustration leads naturally to a scaling approach for understanding the low-temperature spin glass phase. This approach, which has a long history in the study of phase transitions and broken symmetry, examines how the "stiffness" of the low-temperature phase scales with the system size L. A stable phase requires the stiffness — roughly speaking, the free energy cost associated with overturning a droplet of spins — to increase (or at least not decrease) with L, usually as a power law, although other forms are possible in principle.

This approach as applied to spin glasses began with the early work of Anderson and Pond,[29] and was developed throughout the 1980s.[30–38] The essential idea is to study the fluctuations in a finite volume of the spin glass free energy as one changes boundary conditions, for example from periodic to antiperiodic. Physically, such a change in boundary conditions generates relative interfaces inside the box, so one is effectively studying the interface free energy. Given the close relation between this approach and that of the Anderson definition of frustration, one might expect that the presence of frustration will generate profound effects on such interfacial free energies. And of course it does.

4.1. *Interface geometry*

Before proceeding, some remarks are necessary concerning the relation between this procedure and the presence of many states. While switching from periodic to antiperiodic boundary conditions always generates relative interfaces, geometrically one of three things can happen. Consider a "window"[39,40] of large but fixed linear size w centered at the origin, and consider the interfaces generated when $L \gg w$. The first possibility is that as L grows increasingly larger (with w fixed), the interfaces eventually move outside of the window, so that the thermodynamic state *inside* the window is the same for both periodic and antiperiodic boundary conditions. If this is the case, then there are only two spin glass pure states (or ground states at zero temperature), which are global flips of each other.

The other possibility, of course, is that no matter how far away the boundaries move, interfaces always penetrate inside the window. This is the signature of multiple spin glass pure state pairs. This possibility can be further divided into two parts: either the interfaces have vanishing density as the window size increases (with the order of limits being $L \to \infty$ followed by $w \to \infty$) or else the interface density remains bounded away from zero. The former zero-density case is what one finds in the ferromagnet, which exhibits $(d-1)$-dimensional interfaces in a d-dimensional system. In contrast, the latter "space-filling" case, if it occurs, requires d-dimensional interfaces within a d-dimensional system, which would signify a novel feature of spin glasses. Huse and Fisher[41,42] refer to the zero-density situation as "regional congruence": the states are locally the same almost everywhere. The more interesting situation with space-filling interfaces was denoted "incongruence": the states, although similar in a statistical sense, are dissimilar everywhere. It was proven in Ref. 43 that any procedure using boundary conditions chosen in a coupling-independent manner (as in the periodic–antiperiodic situation above) always results either in a single pair of states (no interfaces in the window)[d] or else many incongruent pairs of states. Regionally congruent states, should they exist, can only be generated using coupling-*dependent* boundary conditions, requiring procedures as yet unknown. In what follows we therefore confine the discussion to incongruent states.

---

[d]An absence of interfaces within the window implies that all interfaces necessarily generated by changing boundary conditions must be zero-density, since positive density interfaces *must* penetrate the window.[44] The difference between this case and the regionally congruent case is that in the latter, zero-density interfaces continue to penetrate the window no matter how far away the boundaries are, while in the former, the interfaces deflect to infinity as the boundaries move out to infinity.

4.2. *Interface energetics*

The preceding discussion focuses exclusively on the geometry of interfaces between pure states; we now discuss their energetics. In a ferromagnet, whether homogeneous or random, all couplings have the same sign, so the interface energy scales with the number of edges it comprises. In a spin glass, interface energetics remain an open problem, and a very important one: if one knows how the interface energy scales with its size, one can finally resolve the longstanding open question regarding the multiplicity of pure states in the spin glass phase.

Here's why. By the same reasoning that led to the Anderson definition of frustration, it is reasonable to expect that the free energy of a space-filling interface of linear extent L should scale no faster than $L^{d/2}$. However, it is possible that correlations could lower the minimal interface free energy, so that it scales as L^{θ}, with $0 \leq \theta \leq d/2$ (assuming a stable spin glass phase). The lower bound of $\theta = 0$ is predicted[e] by the mean-field replica symmetry breaking (RSB) picture of the spin glass phase,[22-26] while $0 < \theta < (d-1)/2$ is predicted by the chaotic pairs picture.[40,45,46] Both are many-states pictures, though with very different thermodynamics and organization of the incongruent pure states, as will be discussed below.

Based on numerical results and scaling arguments, Fisher and Huse[35] conjectured a stronger upper bound than $L^{d/2}$; they argued that in fact $\theta \leq (d-1)/2$. If this is correct, then the question of whether incongruent states exist reduces to the question of whether their interface free energy should scale faster or slower than $L^{(d-1)/2}$. Fisher and Huse argued, along roughly similar lines to Anderson, that it should scale as $L^{d/2}$. The resulting contradiction between the upper and lower bounds leads to the conclusion that incongruent states cannot exist in the EA spin glass in any finite dimension.

But is the conjectured upper bound $\theta = (d-1)/2$ correct? This was initially a matter of some controversy, but within a few years it was proved, using rigorous mathematical arguments, by Aizenman and Fisher,[47] and independently and a little later by Newman and Stein.[48] Unfortunately, neither was ever published, but the arguments are now familiar to those

---

[e]More precisely, this scaling applies to interfaces between pairs of incongruent states within the *same* thermodynamic state. Interfaces between incongruent states belonging to *different* thermodynamic states, which would be expected within the RSB picture upon switching from periodic to antiperiodic boundary conditions, would presumably have $\theta > 0$. For a discussion of thermodynamic states within RSB, see Ref. 46, Sec. 7.9.

who work in this area. I will informally sketch the basic idea of the proof here. First, let me state the exact result more formally:

Theorem 1. *Let F_P be the free energy of the finite-volume Gibbs state generated by Hamiltonian (2) (with $h = 0$) in a box Λ of volume L^d using periodic boundary conditions, and F_{AP} that generated using antiperiodic boundary conditions. Let $X_\Lambda = F_P - F_{AP}$. Then $\mathrm{Var}(X_\Lambda) \le \mathrm{const.} \times L^{d-1}$, where $\mathrm{Var}(\cdot)$ denotes the variance over all of the couplings inside the box.*

Physically, this implies that the fluctuations in free energy (and therefore the interface free energy) induced by changing boundary conditions from periodic to antiperiodic in a box of volume L^d scales as $L^{(d-1)/2}$. The theorem as stated above is more restrictive than necessary; the same result applies for any two boundary conditions that are *gauge-related*, i.e., that can be transformed into each other by reversing the sign of some subset of couplings on the boundary of the box.[f] So, for example, the same result holds for any two distinct fixed boundary conditions.

The proof uses a *martingale decomposition* of the free energy difference, as follows. First note that, by the gauge-relatedness of the two boundary conditions, $E[X_\Lambda] = 0$, where $E[\cdot]$ denotes a full average over all the couplings inside the box. Next number each coupling inside the box, as shown in Fig. 1. We now successively average the free energy difference X_Λ over an increasing number of couplings. Let $x_{\Lambda,j} = E[X_\Lambda | b_1 b_2 \ldots b_j]$, where $E[A | y_1, y_2, \ldots, y_k]$ is the conditional expectation, or average, of A conditioned on the random variables $y_1, y_2, \ldots, y_k$. That is, suppose that A is a random quantity depending on the N random variables $y_1, y_2, \ldots, y_N$. Then $E[A | y_1, y_2, \ldots, y_k]$ represents the quantity resulting from averaging A over $y_{k+1}, y_{k+2}, \ldots, y_N$. So $x_{\Lambda,0}$ represents X_Λ fully averaged over all of the couplings in the interior of the box, and $x_{\Lambda,N} = X_\Lambda$, the original unaveraged free energy difference.

If there are N couplings inside the box, it is not hard to see that

$$\mathrm{Var}(X_\Lambda) = \mathrm{Var}\left[\sum_{j=0}^{N-1}(x_{\Lambda,j+1} - x_{\Lambda,j})\right] = \sum_{j=0}^{N-1} \mathrm{Var}(x_{\Lambda,j+1} - x_{\Lambda,j}), \quad (6)$$

where the second equality follows because the so-called martingale differences $x_{\Lambda,j+1} - x_{\Lambda,j}$ are orthogonal quantities.[g]

---

[f] A coupling on the boundary of the box connects a site inside Λ to one on the boundary of Λ.

[g] Two random variables A and B are orthogonal if $E[AB] = 0$. It is easy to see that this holds for any two martingale differences.

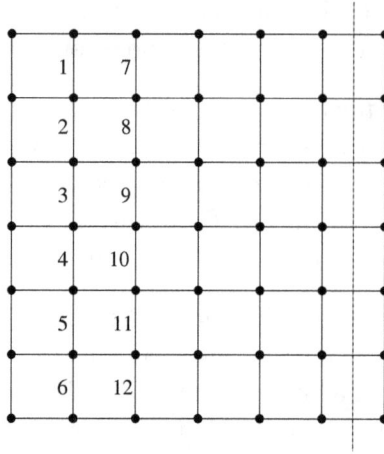

Fig. 1.   Schematic of a box with each coupling numbered (only the first twelve are explicitly indicated). The dashed line on the right indicates couplings whose transformation
$J_{xy} \rightarrow -J_{xy}$ is equivalent to switching from periodic to antiperiodic boundary conditions,
and which according to the argument in the text are the only ones contributing to the
variance.

Now suppose that the antiperiodic boundary conditions are applied to the
right and left boundaries, and consider an $x_{\Lambda,j}$ conditioned on any subset
of the couplings *except* those cut by the dashed line in Fig. 1. Any such
$x_{\Lambda,j} = 0$, because one is averaging over all of the couplings cut by the dashed
line, and taking $J_{xy} \rightarrow -J_{xy}$ over these couplings is equivalent to switching
between periodic and antiperiodic boundary conditions. The only nonzero
contributions to the sum in (6) therefore comes from conditioning on these
boundary couplings, each of which contributes a term of order one to the
variance. This completes the argument.

5. Interface Free Energy Fluctuations

If we can now find a strong *lower* bound for interface free energies between
incongruent states, we would be in a position to determine whether such
states can exist at all; and if they do, what their properties should be. Unfortunately, finding a lower bound is considerably more difficult than finding
an upper bound, for reasons that will be discussed below. It has been almost
25 years since the upper bound was proved, and no progress on finding a
lower bound has been made until very recently.

Unlike the upper bound, which is scenario-independent, construction of
a lower bound requires a specific picture of the spin glass phase. In other

words, one first needs to ask: what's doing the fluctuating? We will examine here four distinct pictures that have been proposed, each of which gives a different answer to the question.

5.1. *Scenarios for the low-temperature spin glass phase*

Probably the most familiar are the *mixed-state* pictures, in which the spin glass phase consists of infinitely many thermodynamic states, each of which is itself a mixture of infinitely many incongruent pure state pairs with nonzero weights (within their respective thermodynamic states). The mean-field-inspired RSB scenario is such a picture, although others are also possible. Recall from Sec. 4.2 that in these scenarios the smallest interface free energies remain order one independently of the interface size ($\theta = 0$).

Almost equally familiar are the scaling/droplet pictures discussed in Sec. 4. These are two-state pictures with $\theta > 0$ and in which no interface appears in a window far from the boundaries when one switches from periodic to antiperiodic boundary conditions.

There are two other less familiar pictures that should nevertheless be included: the "TNT" picture of Krzakala-Martin[49] and Palassini–Young,[50] and the chaotic pairs picture discussed in Sec. 4.2. If one constructs a 2×2 grid listing the different possibilities for interface geometry (space-filling vs. zero-density) and energetics ($\theta = 0$ vs. $\theta > 0$) then these additional pictures are required for completeness.[51,h] The TNT picture is presumably a two-state picture[43] with zero-density interfaces and $\theta = 0$, while chaotic pairs is a many-state picture with space-filling interfaces and $\theta > 0$. Unlike RSB, it is not a (non-trivial) mixed-state picture: while it contains infinitely many distinct thermodynamic states, each one consists of a single spin-reversed pure state pair.

So of these four, only RSB and chaotic pairs possess incongruent states, and any lower bound on interface free energies can help to determine whether they are likely candidates for the spin glass phase, or else are not allowed at all. How so? Consider again the energetics of interfaces between putative incongruent pure states. As already mentioned, unlike in the ferromagnet, whether homogeneous, random-bond, or random-field, the sign of the free energy difference between two putative spin glass states varies as one moves along their relative interface. If the single-coupling energy differences are independent, then one expects an energy that varies as the square root of

---

[h]They also arise naturally from a metastate analysis of possible spin glass phases.[40,45,52,53]

the number of couplings in the interface — i.e., in a volume of size L, the fluctuations in the interface free energy would scale as $L^{d/2}$, as in the conjecture accompanying the Anderson definition of frustration. If this were indeed the case, such fluctuations would violate, in any finite dimension, the upper bound of $L^{(d-1)/2}$ described in Sec. 4.2, and the existence of incongruent states would therefore be impossible in finite-dimensional spin glasses.

But the single-coupling free energy differences are certainly *not* independent. Whether the correlations in free energy fluctuations as one moves along the interface are strong enough to decrease the free energy scaling exponent from its independent value is the crucial question that determines whether incongruent states are present in spin glasses — or not. Hence, determining an accurate lower bound is of central importance for resolving the question of multiplicity of states in realistic spin glasses.

5.2. *Lower bound*

If it's so important, why has it taken so long to find a lower bound for free energy difference fluctuations? Progress has been held back by several technical hurdles; the two most troublesome are the "cancellation problem" and the "identification problem".

The cancellation problem has already been discussed in Sec. 4.2. If incongruent states exist, then an arbitrarily chosen coupling will have, with positive probability, a free energy difference of order one between the two states. But if one uses the usual techniques, such as martingale differences, to extrapolate to the entire volume, cancellations between the many terms lead to an ambiguous outcome.

Equally difficult is the identification problem. It is hard to see how one can estimate the extent of free energy difference fluctuations between two thermodynamic states Γ and Γ' without averaging over the couplings inside the volume. But as one does so, what happens to the original states? Unlike in ferromagnets and other homogeneous systems, there is no clearcut connection between boundary conditions and thermodynamic states, and (if there are many incongruent states) the states themselves can change as one varies the couplings inside the box. So *a priori* it is not even clear what one is calculating during the averaging procedure.

In a very recent paper,[54] these and other problems were finally surmounted, although (at the moment) for a limited class of incongruent states. For these incongruent states it was found that the fluctuations in free energy differences do indeed scale as $L^{d/2}$; or more formally, the variance of the free energy difference between the incongruent states considered scales linearly with the volume.

If these results can be extended to the set of incongruent states in general, does one then have a contradiction? In a strict mathematical sense, not quite yet.[i] The problem is that the quantity for which the lower bound is derived is not exactly the same as that for which the upper bound was derived, although both are just different representations of the free energy difference in a finite volume, and so are equivalent in a physical sense.

The upshot is that we may be on the cusp of resolving the problem of multiplicity of pure states in realistic spin glasses — but at the moment it is unclear as to whether the results can be extended to bring the upper and lower bounds into alignment. Whether this is eventually done or not, it is clear that the insights that Phil Anderson had 35 years ago into the nature of frustration are still actively guiding and influencing fundamental research today.

Acknowledgments

The author thanks Louis-Pierre Arguin and Chuck Newman for useful comments on the manuscript, and Louis-Pierre Arguin, Chuck Newman, and Janek Wehr for an interesting and enjoyable collaboration that led to the work described in Sec. 5.2. This research has been supported in part by U.S. National Science Foundation Grants DMS 1207678 and OISE 0730136.

References

1. P.W. Anderson, Plasmons, gauge invariance, and mass, *Phys. Rev.* **130**, 439–442 (1962).
2. P.W. Anderson, More is different, *Science* **177**, 393–396 (1972).
3. P.W. Anderson. Spin glass Hamiltonians: A bridge between biology, statistical mechanics and computer science, in D. Pines (Ed.), *Emerging Syntheses in Science* (Addison-Wesley, 1988), pp. 17–20.
4. P.W. Anderson, Lectures on amorphous systems, in R. Balian, R. Maynard and R. Toulouse (Eds.), *La Matiére mal condensèe (Ill-condensed matter)* (North-Holland, 1979), pp. 159–261.
5. S. Edwards and P.W. Anderson, Theory of spin glasses, *J. Phys. F* **5**, 965–974 (1975).

---

[i]Whether such a result already provides a sufficient heuristic or theoretical physics-style argument for the non-existence of incongruent states is left up to the reader.

6. D. Sherrington and S. Kirkpatrick, Solvable model of a spin glass, *Phys. Rev. Lett.* **35**, 1792–1796 (1975).

7. G. Toulouse, Theory of frustration effect in spin-glasses, *Commun. Phys.* **2**, 115 (1977).

8. P.W. Anderson, The concept of frustration in spin glasses, *J. Less Common Metals* **62**, 291–294 (1978).

9. P.W. Anderson. Survey of theories of spin glass, in R.A. Levy and R. Hasegawa (Eds.), *Amorphous Magnetism II* (Plenum Press, 1977), pp. 1–16.

10. P.W. Anderson, Spin glass I: A scaling law rescued, *Physics Today* **41(1)**, 9–11 (1988).

11. P.W. Anderson, Spin glass II: Is there a phase transition? *Physics Today* **41(3)**, 9 (1988).

12. P.W. Anderson, Spin glass III: Theory raises its head, *Physics Today* **41(6)**, 9–11 (1988).

13. P.W. Anderson, Spin glass IV: Glimmerings of trouble, *Physics Today* **41(9)**, 9–11 (1988).

14. P.W. Anderson, Spin glass V: Real power brought to bear, *Physics Today* **42(7)**, 9–11 (1989).

15. P.W. Anderson, Spin glass VI: Spin glass as cornucopia, *Physics Today* **42(9)**, 9–11 (1989).

16. P.W. Anderson, Spin glass VII: Spin glass as paradigm, *Physics Today* **43(3)**, 9–11 (1990).

17. Y. Imry and S.-K. Ma, Random-field instability of the ordered state of continuous symmetry, *Phys. Rev. Lett.* **35**, 1399–1401 (1975).

18. G. Parisi and N. Sourlas, Random magnetic fields, supersymmetry, and negative dimensions, *Phys. Rev. Lett.* **43**, 744–745 (1979).

19. J. Imbrie, The ground state of the three-dimensional random-field Ising model, *Commun. Math. Phys.* **98**, 145–176 (1985).

20. M. Aizenman and J. Wehr, Rounding effects of quenched randomness on first-order phase transitions, *Commun. Math. Phys.* **130**, 489–528 (1990).

21. J. Wehr and M. Aizenman, Fluctuations of extensive functions of quenched random couplings, *J. Stat. Phys.* **60**, 287–306 (1990).

22. G. Parisi, Infinite number of order parameters for spin-glasses, *Phys. Rev. Lett.* **43**, 1754–1756 (1979).

23. G. Parisi, Order parameter for spin-glasses, *Phys. Rev. Lett.* **50**, 1946–1948 (1983).

24. M. Mézard, G. Parisi, N. Sourlas, G. Toulouse and M. Virasoro, Nature of spin-glass phase, *Phys. Rev. Lett.* **52**, 1156–1159 (1984).

25. M. Mézard, G. Parisi, N. Sourlas, G. Toulouse and M. Virasoro, Replica symmetry breaking and the nature of the spin-glass phase, *J. Phys. (Paris).* **45**, 843–854 (1984).

26. M. Mézard, G. Parisi and M.A. Virasoro, Eds., *Spin Glass Theory and Beyond* (World Scientific, 1987).

27. E. Fradkin, B.A. Huberman and S.H. Shenker, Gauge symmetries in random magnetic systems, *Phys. Rev. B* **18**, 4789–4814 (1978).

28. Bovier and J. Fröhlich, A heuristic theory of the spin glass phase, *J. Stat. Phys.* **44**, 347–391 (1986).

29. P.W. Anderson and C.M. Pond, Anomalous dimensionalities in the spin-glass problem, *Phys. Rev. Lett.* **40**, 903–906 (1978).

30. J.R. Banavar and M. Cieplak, Nature of ordering in spin-glasses, *Phys. Rev. Lett.* **48**, 832–835 (1982).

31. J.R. Banavar and M. Cieplak, Scaling stiffness of spin glasses, *J. Phys. C* **16**, L755–L759 (1983).

32. W.L. McMillan, Scaling theory of Ising spin glasses, *J. Phys. C* **17**, 3179–3187 (1984).

33. R.G. Caflisch and J.R. Banavar, Renormalization-group study of interfacial properties and its applications to an Ising spin glass, *Phys. Rev. B* **32**, 7617–7620 (1985).

34. A.J. Bray and M.A. Moore, Critical behavior of the three-dimensional Ising spin glass, *Phys. Rev. B* **31**, 631–633 (1985).

35. D.S. Fisher and D.A. Huse, Ordered phase of short-range Ising spin-glasses, *Phys. Rev. Lett.* **56**, 1601–1604 (1986).

36. A.J. Bray and M.A. Moore, Chaotic nature of the spin-glass phase, *Phys. Rev. Lett.* **58**, 57–60 (1987).

37. D.S. Fisher and D.A. Huse, Nonequilibrium dynamics of spin glasses, *Phys. Rev. B* **38**, 373–385 (1988).

38. D.S. Fisher and D.A. Huse, Equilibrium behavior of the spin-glass ordered phase, *Phys. Rev. B* **38**, 386–411 (1988).

39. C.M. Newman and D.L. Stein, Simplicity of state and overlap structure in finite-volume realistic spin glasses, *Phys. Rev. E* **57**, 1356–1366 (1998).

40. C.M. Newman and D.L. Stein, Topical Review: Ordering and broken symmetry in short-ranged spin glasses, *J. Phys.: Condens. Matter* **15**, R1319–R1364 (2003).

41. D.A. Huse and D.S. Fisher, Pure states in spin glasses, *J. Phys. A* **20**, L997–L1003 (1987).

42. D.S. Fisher and D.A. Huse, Absence of many states in realistic spin glasses, *J. Phys. A* **20**, L1005–L1010 (1987).

43. C.M. Newman and D.L. Stein, Interfaces and the question of regional congruence in spin glasses, *Phys. Rev. Lett.* **87**, 077201-1-077201-4 (2001).

44. C.M. Newman and D.L. Stein, The state(s) of replica symmetry breaking: Mean field theories vs. short-ranged spin glasses, *J. Stat. Phys.* **106**, 213–244 (2002).

45. C.M. Newman and D.L. Stein, Spatial inhomogeneity and thermodynamic chaos, *Phys. Rev. Lett.* **76**, 4821–4824 (1996).

46. D.L. Stein and C.M. Newman, *Spin Glasses and Complexity* (Princeton University Press, 2013).

47. M. Aizenman and D.S. Fisher, unpublished.

48. C.M. Newman and D.L. Stein, unpublished.

49. F. Krzakala and O.C. Martin, Spin and link overlaps in three-dimensional spin glasses, *Phys. Rev. Lett.* **85**, 3013–3016 (2000).

50. M. Palassini and A.P. Young, Nature of the spin glass state, *Phys. Rev. Lett.* **85**, 3017–3020 (2000).

51. C.M. Newman and D.L. Stein, Finite-dimensional spin glasses: States, excitations, and interfaces, *Ann. Henri Poincaré, Suppl. 1.* **4**, S497–S503 (2003).

52. C.M. Newman and D.L. Stein, Metastate approach to thermodynamic chaos, *Phys. Rev. E.* **55**, 5194–5211 (1997).

53. C.M. Newman and D.L. Stein. Thermodynamic chaos and the structure of short-range spin glasses, in A. Bovier and P. Picco (Eds.), *Mathematics of Spin Glasses and Neural Networks* (Birkhauser, 1998), pp. 243–287.

54. L.-P. Arguin, C.M. Newman, D.L. Stein and J. Wehr, Fluctuation bounds for interface free energies in spin glasses, *J. Stat. Phys.* (2014), published online May 12, 2014.

Phil Anderson's Magnetic Ideas in Science

Piers Coleman

Center for Materials Theory, Department of Physics and Astronomy
Rutgers University, Piscataway, NJ 08854, USA

and

Department of Physics, Royal Holloway, University of London
Egham, Surrey TW20 0EX, UK
coleman@physics.rutgers.edu

In Philip W. Anderson's research, magnetism has always played a special role, providing a prism through which other more complex forms of collective behavior and broken symmetry could be examined. I discuss his work on magnetism from the 1950s, where his early work on antiferromagnetism led to the pseudospin treatment of superconductivity — to the 1970s and 1980s, highlighting his contribution to the physics of local magnetic moments. Phil's interest in the mechanism of moment formation, and screening evolved into the modern theory of the Kondo effect and heavy fermions.

1. Introduction

This article is based on a talk I gave about Phil Anderson's contributions to our understanding of magnetism and its links with superconductivity, at the 110th Rutgers Statistical mechanics meeting. This event, organized by Joel Lebowitz, was a continuation of the New Jersey celebrations began at "PWA90: A lifetime of emergence", on the weekend of Phil Anderson's 90th birthday in December 2013. My title has a double-entendre, for Phil's ideas in science have a magnetic quality, and have long provided inspiration, attracting students such as myself, to work with him. I first learned about Phil Anderson as an undergraduate at Cambridge in 1979, some three years after he had left for Princeton. Phil had left behind many legends at Cambridge, one of which was that he had ideas of depth and great beauty, but also that he was very hard to understand. For me, as with many fellow students of Phil, the thought of working with an advisor with some of the best ideas on the block was very attractive, and it was *this* magnetism that brought me over to New York Harbor nine months later, to start a PhD with Phil at Princeton.

One of the recurrent themes of Anderson's work, is the importance of using models as a gateway to discovering general mechanisms and principles, and throughout his career, models of magnetism played a key role. In his book *Basic Concepts of Condensed Matter Physics*,[1] Anderson gives various examples of such basic principles, such as adiabatic continuation, the idea of renormalization as a way to eliminate all but the essential degrees of freedom, and most famously, the link between broken symmetry and the idea of *generalized rigidity*, writing

> "We are so accustomed to the rigidity of solid bodies that is hard to realize that such action at a distance is not built into the laws of nature. It is strictly a consequence of the fact that the energy is minimized when symmetry is broken in the same way throughout the sample.
>
> The generalization of this concept to all instances of broken symmetry is what I call generalized rigidity. It is responsible for most of the unique properties of the ordered (broken-symmetry) states: ferromagnetism, superconductivity, superfluidity, etc."

Yet in the 1950s, when Phil began working on magnetism, these ideas had not yet been formed: the term *broken symmetry* was not yet in common usage, renormalization was little more than a method of eliminating divergences in particle physics and beyond the Ising and Heisenberg models, there were almost no other simple models for interacting electrons. Phil's studies of models of magnetism spanning the next three decades played a central role in the development of his thoughts on general principles and mechanisms in condensed matter physics, especially those underlying broken symmetry.

I'll discuss three main periods in Phil's work as shown in the time-line of Fig. 1, and arbitrarily color coded as the "blue", "orange" and "green" period. My short presentation is unfortunately highly selective but I hope it will give a useful flavor to the reader of the evolution of ideas that have accompanied Phil's work in magnetism.

2. Blue Period: Antiferromagnetism and Superconductivity

Today it is hard to imagine the uncertainties connected with antiferromagnetism and broken symmetry around 1950. While Néel and Landau[2,3] had

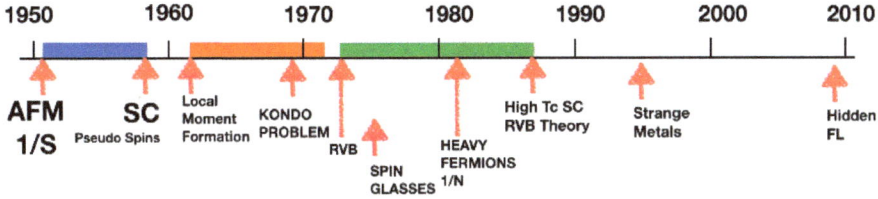

Fig. 1.   Three periods of Anderson's research into magnetism selectively discussed in this article. Blue period: from antiferromagnetism to superconductivity. Orange period: theory of local moment formation and the Kondo problem. Green period: from resonating valence bonds (RVB) to high-temperature superconductivity.

independently predicted "antiferromagnetism", with a staggered magnetization (↑↓↑↓↑↓), as the classical ground-state of the Heisenberg model with positive exchange interaction,

$$H = J \sum_{(i,j)} \vec{S}_i \cdot \vec{S}_j, \qquad (J > 0), \tag{1}$$

the effects of quantum fluctuations were poorly understood. Most notably, the one-dimensional $S = 1/2$ model had been solved exactly by Bethe in 1931,[4] and in his *Bethe Ansatz* solution, it was clear there was no long range order, indicating that at least in one dimension, quantum fluctuations overcome the long-range order. This issue worried Landau so much, that by the 1940s he had abandoned the idea of antiferromagnetism in quantum spin systems.[5] Phil Anderson reflects on this uncertainty in his 1952 article "An Approximate Quantum Theory of the Antiferromagnetic Ground State",[6] writing

> "For this reason the very basis for the recent theoretical work which has treated antiferromagnetism similarly to ferromagnetism remains in question. In particular, since the Bethe–Hulthén ground-state is not ordered, it has not been certain whether an ordered state was possible on the basis of simple $\vec{S}_i \cdot \vec{S}_j$ interactions"

The situation began to change in 1949, with Shull and Smart's[7] detection of antiferromagnetic order in MnO by neutron diffraction, which encouraged Anderson to turn to the unsolved problem of zero-point motion in antiferromagnets. Early work on spin-wave theory by Heller, Kramers and Hulthén had treated spin waves as classical excitations, but later work by Klein and Smith[8] had noted that quantum zero-point motions in a spin-S ferromagnet correct the ground-state energy by an amount of order $1/S$, a quantity that becomes increasingly small as the size of the spin increases. It is this effect that increases the ground-state energy of a ferromagnet from its classical value $E \propto -J \langle \vec{S}^2 \rangle = -JS(S+1)$ to its exact quantum value $E \propto -JS^2$.

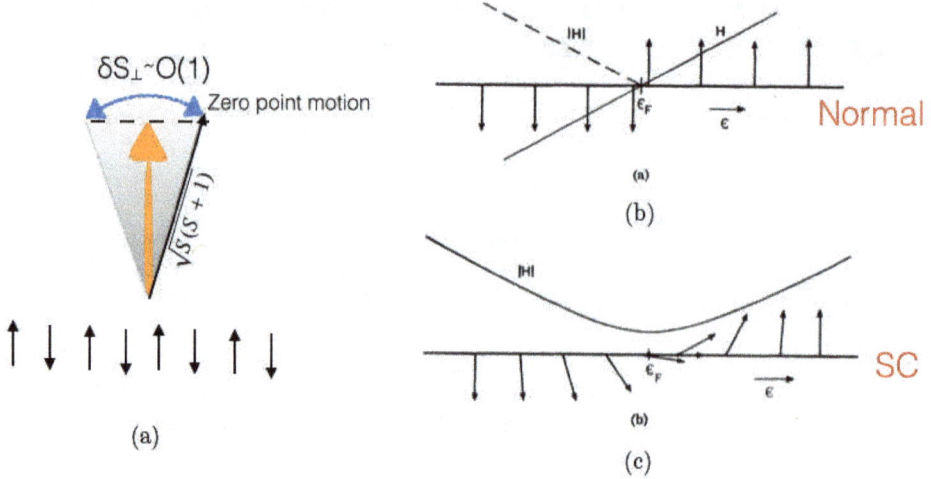

Fig. 2. (a) Reduction of staggered moment from length $\sqrt{S(S+1)}$ to semiclassical value $\sqrt{S(S+1)} - O(1)$. (b) In Anderson's domain wall interpretation of superconductivity the normal Fermi liquid is a sharp domain wall where the Weiss field H vanishes at the Fermi surface; (c) in the superconductor the pseudo-spins rotate smoothly and Weiss field never vanishes, giving rise to a finite gap.

A key result of Anderson's work is an explanation for the survival of antiferromagnetic order in two and higher dimensions, despite its absence in the Bethe chain. His expression for the reduced sublattice magnetization (Fig. 2a) of a bipartite antiferromagnet, is

$$\langle S_z \rangle = \sqrt{S(S+1) - \delta S_\perp^2}$$

$$= S \left[1 - \frac{1}{2S} \int \frac{d^d q}{(2\pi)^d} \left(\frac{1}{\sqrt{1 - \gamma_q^2}} - 1 \right) + O\left(\frac{1}{S^2} \right) \right] \quad (2)$$

where $\gamma_q = \frac{1}{d} \sum_{l=1,d} \cos q_l$. A similar result was independently discovered by Ryogo Kubo.[9] In an antiferromagnet, the staggered magnetization does not commute with the Hamiltonian and thus undergoes continuous zero-point fluctuations that reduce its magnitude (Fig. 2a). Since $\sqrt{1 - \gamma_k^2} \sim |q|$ at small wave vector q, these fluctuations become particularly intense at long wavelengths, with a reduction in magnetization

$$\Delta M \sim \int \frac{d^d q}{q} \sim \begin{cases} \infty & (d = 1) \\ \text{finite} & (d \geq 2) \end{cases} . \quad (3)$$

In this way, Anderson's model calculation could account for the absence of long range order in the Bethe chain as a result of long-wavelength quantum fluctuations and the stability of antiferromagnetism in higher dimensions.

At several points in his paper, Phil muses on the paradox that the ground-state of an antiferromagnet is a singlet, with no preferred direction, a thought he would return to in his later work on resonating valence bonds with Patrick Fazekas.[10] For the moment however, Phil resolves the paradox by estimating that the time for an antiferromagnet to invert its spins by tunnelling is macroscopically long, so that the sublattice magnetization becomes an observable classical quantity. Phil's semiclassical treatment of the antiferromagnet would later set the stage for Duncan Haldane[11] to carry out a semiclassical treatment of the one-dimensional Heisenberg model, revealing an unexpected topological term. But in the near future, Anderson's study of antiferromagnetism had influence in a wholly unexpected direction: superconductivity.

One of the issues that was poorly understood following the Bardeen–Cooper–Schrieffer (BCS) theory of superconductivity, was the question of charge fluctuations. In an insulator the charge gap leads to a dielectric with no screening. However, were the superconducting gap to have the same effect, it would eliminate the weak electron-phonon attraction. It was thus essential to show that both screening and the longitudinal plasma mode survive the formation of the BCS gap. In his 1958 paper "Random-Phase Approximation in the Theory of Superconductivity",[12] Phil writes

> "Both for this reason, and because it seems optimistic to assume that the collective [charge fluctuation] and screening effects (which are vital even in determining the phonon spectrum) will be necessarily unaffected by the radical changes in the Fermi sea . . . , it is desirable to have a theory of the ground-state of a superconductor which can simultaneously handle these collective effects in the best available approximation . . ."

Phil's experience with antiferromagnetism enabled him to make a new link between magnetism and superconductivity. He observed that if one considered a pair to be a kind of "down-spin" and the absence of a pair to be a kind of "up-spin" in particle hole space,

$$\text{no pair:} \qquad |\Uparrow\rangle \equiv |0\rangle,$$

$$\text{pair:} \qquad |\Downarrow\rangle \equiv c^\dagger_{k\uparrow} c^\dagger_{-k\downarrow} |0\rangle \qquad (4)$$

then the BCS ground-state is revealed as a kind of Bloch domain wall (Fig. 2b and 2c) formed around the Fermi surface.[12] This new interpretation forges a link between between superconductivity and antiferromagnetism, enabling the pairing field to be identified as a transverse Weiss field in particle-hole space. Moreover the analogy works at a deeper level, because like in quantum antiferromagnetism, the superconducting order parameter is non-conserved, allowing it to fluctuate and importantly, to deform in response to an electric field, preserving the screening.

Let us look at this in a little more detail. BCS theory involves three key operators, the number operator $(n_{k\uparrow} + n_{-k\downarrow})$, the pair creation and pair annihilation operators, $b_k^\dagger = c_{k\uparrow}^\dagger c_{-k\downarrow}^\dagger$, $b_k = c_{-k\downarrow} c_{k\uparrow}$. The key observation was to identify these operators as the components of a pseudo-spin. In the subspace where $n_{k\uparrow} + n_{-k\downarrow}$ is either 0 or 2, Anderson's defined the pseudo-spin as

$$2s_z = 1 - n_k - n_{-k} = \begin{matrix} \text{empty} \\ \text{full} \end{matrix} \begin{bmatrix} \text{empty} & \text{full} \\ 1 & 0 \\ 0 & -1 \end{bmatrix}, \tag{5}$$

so that a fully occupied k state is a "down" pseudo-spin, and an empty k-state is an "up" spin. Similarly, the raising and lowering operators are respectively, the pair destruction and creation operators

$$b_k \equiv s_{xk} + is_{yk} = \begin{pmatrix} 0 & 1 \\ 0 & 0 \end{pmatrix}, \qquad b_k^\dagger \equiv s_{xk} - is_{yk} = \begin{pmatrix} 0 & 0 \\ 1 & 0 \end{pmatrix}. \tag{6}$$

In this language, the BCS reduced Hamiltonian

$$\mathcal{H}_{\text{RED}} = -2 \sum_k \epsilon_k s_{zk} - \sum_{k,k'} V_{k,k'} \vec{s}_{\perp k} \cdot \vec{s}_{\perp k'} \tag{7}$$

is a kind of magnet that resides in momentum space. Anderson showed that in this language, the metal is a sharp domain wall along the Fermi surface (Fig. 2b) while the superconductor has a soft "Bloch domain wall" (Fig. 2c) in which the pseudo-spins rotate continuously from down (full) to sideways (linear combination of full and empty) to up (empty). By calculating the spin-wave fluctuations Anderson was able to show that with the Coulomb interaction included, the longitudinal electromagnetism of the metal, its screening and plasma modes, are unaffected by the superconducting gap.

Anderson's pseudo-spin reformulation of BCS had a wide influence. Two years later, Nambu extended the pseudo-spin approach to reformulate Gor'kov's Green function approach using his now famous "Nambu matrices".[13] Perhaps most important of all, by making the analogy between superconductivity and magnetism, the community took a cautious step closer to regarding the superconducting phase as a palpable, detectable variable (with the caveats of gauge fixing). This new perspective, especially the link between phase, supercurrents and gauge invariance, would soon culminate in Anderson's ideas on how gauge particles acquire mass — the *Anderson Higgs* mechanism (see Witten's article in this volume).

3. Orange Period: Moment Formation and the Kondo Problem

3.1. *Superexchange*

Towards the end of the 1950s, Anderson began to turn his attention towards the microscopic origins of antiferromagnetism. In his 1959 paper "New Approach to the Theory of Superexchange Interactions"[14] Phil argues that origin of antiferromagnetism is *the Mott mechanism*, i.e. the Coulomb cost of doubly occupied orbitals. Phil writes

> "In such a simple model all the degenerate states in the ground-state manifold have exactly one electron per ion, while all the excited states with one transferred electron have energy U. Between an pair of ions at a distance $\mathbf{R} - \mathbf{R}'$ there is only one (hopping term) $b_{\mathbf{R}-\mathbf{R}'}$; this must act to return the state to one of the ground manifold... so that

$$\Delta E = \text{constant} + \sum_{\mathbf{R},\mathbf{R}'} \frac{2|b_{\mathbf{R}-\mathbf{R}'}|^2}{U} \mathbf{S_R} \cdot \mathbf{S'_{R'}}, \qquad (8)$$

> This is the antiferromagnetic exchange effect."

This paper contains the origins of our modern understanding of Mott insulators, including an early formulation of the Hubbard model with Anderson's hallmark use of U to denote the onsite Coulomb repulsion.

3.2. *Anderson's model for local moment formation*

While the notion of local moments is rooted in early quantum mechanics, the mechanism of moment formation was still unknown in the 1950s. At this time, experiments at Orsay at Bell Labs started to provide valuable new insights. In Orsay, Jacques Friedel and André Blandin proposed that virtual bound-states develop around localized d-states in a metal, arguing that ferromagnetic exchange forces then split these resonances to form local moments. Recalling the first time he encountered this idea, Phil writes[15,16]:

> "In the Fall of '59, a delightful little discussion meeting on magnetism in metals was held in Brasenose College, Oxford. ...Blandin presented the idea of virtual states and I the conceptual basis for antiferromagnetic s-d exchange, without any understanding, at least on my part, that the two ideas belonged together. The only immediate positive scientific result of

the meeting was that I won a wager on the sign of the Fe hyperfine field
on the basis of these ideas."

Around this time, Bernd Matthias's group at at Bell Labs discovered that
the development of a localized moment on iron atoms depends on the metallic environment — for example, iron impurities dissolved in niobium do not
develop a local moment, yet they do so in the niobium-molybdenum alloy,
$Nb_{1-x}Mo_x$ once the concentration of molybdenum exceeds 40% $(x > 0.4)$.
Anderson was intrigued by this result and realized that while it was probably
connected to the virtual bound-state ideas of Friedel and Blandin, ferromagnetic exchange was two weak to drive moment formation. Once again, he
turned to the Mott mechanism as a driver and the key element of his theory
"Localized Magnetic States in Metals", is the repulsion between *anti-parallel*
electrons in the same orbital, given by the Coulomb repulsion integral,

$$U = \int |\phi_{\text{loc}}(1)|^2 e^2 r_{12}^{-1} |\phi_{\text{loc}}(2)|^2 d\tau. \tag{9}$$

Phil emphasizes this point, writing

> "the formal theory is much more straightforward if one includes U in the
> manner in which we do it, as a repulsion of opposite-spin electrons in
> ϕ_{loc}, not as an attraction of parallel ones"

Another new element of Anderson's theory of moment formation, not contained in earlier theories, was the explicit formulation of his model as a
quantum field theory. The heart of the Anderson model is a hybridization
term

$$H_{sd} = \sum_{\mathbf{k},\sigma} V_{d\mathbf{k}} (c_{\mathbf{k}\sigma}^{\dagger} c_{d\sigma} + c_{d\sigma}^{\dagger} c_{\mathbf{k}\sigma}), \tag{10}$$

which mixes s and d electrons, generating a virtual bound-state of width
$\Delta = \pi \langle V^2 \rangle \rho$, where $\rho(\epsilon)$ is the conduction electron density of states and
$\langle V^2 \rangle$ the Fermi surface average of the hybridization, and the on-site Coulomb
interaction,

$$H_{\text{corr}} = U n_{d\uparrow} n_{d\downarrow}, \tag{11}$$

where U is as given in (9). With these two terms, Phil unified the Freidel–
Blandin virtual bound-state resonance with the "Mott mechanism" he
had already introduced for insulating antiferromagnets. Using a mean-field
Hartree–Fock treatment of his model, Anderson shows that if

$$U > \pi\Delta \tag{12}$$

the virtual bound-state resonance splits into two. One of the aspects of the
paper that may have been confusing at the time, was that taken literally,

this suggested a real phase transition into a local moment state. Anderson clearly did not see it this way,

> "It is the great conceptual simplification of the impurity problem that is possible to separate the question of the existence of the "magnetic state" entirely from the actually irrelevant question of whether the final state is ferromagnetic, antiferromagnetic or paramagnetic."

Today we understand that Anderson's mean-field description of magnetic moments captures the physics at intermediate time scales, describing a crossover in the renormalization trajectory as it makes a close fly-by past the repulsive local moment captured in Phil's mean field theory.

3.3. *The Kondo model*

A central prediction of Phil's 1961 paper was that the residual interaction between the local moment, and the surrounding electrons, the *s-d* interaction is *antiferromagnetic*. By freezing the local moment, Phil was able to calculate the small shifts in the conduction electron energies, demonstrating they were indeed antiferromagnetic. He writes

> "Thus any *g* shifts caused by free electron polarization will tend to have antiferromagnetic sign."

Phil had of course guessed this in his 1959 bet at Brasenose College Oxford!
    The conventional wisdom of the time expected a ferromagnetic *s-d* exchange. Indeed, the *s-d model* describing the interaction of local moments with conduction electrons had been formulated by Clarence Zener[17] and written down in second-quantized notation in the 1950s, by Tadao Kasuya,[18] but with a *ferromagnetic* interaction, derived from exchange.[a]
    On the other side of the world in Tokyo, one person, Jun Kondo realized that Anderson's prediction of an antiferromagnetic *s-d* coupling would have experimental consequences, and his efforts to reveal them led to the solution of a 30-year-old mystery. Following Anderson's prediction, Kondo now wrote

---

[a]Curiously, in his 1961 paper, Phil does not mention superexchange as the origin of the antiferromagnetic *s-d* interaction, despite his development of this idea in his 1959 paper on insulating antiferromagnets. Perhaps it was felt that metals are different. It was not until the work of Schrieffer and Wolff[19] that the Kondo interaction was definitively identified, using a careful canonical transformation, as a form of superexchange interaction, of magnitude

$$J \sim \frac{4\langle V_{d\mathbf{k}}^2 \rangle}{U}. \tag{13}$$

down a simple model for the *antiferromagnetic* interaction,

$$H_K = \sum \epsilon_k c_{k\sigma}^\dagger c_{k\sigma} + J\vec{S}_d \cdot \vec{\sigma}(0),$$ (14)

where $\vec{\sigma}(0)$ is the electron spin density at the site of the magnetic moment, and he set out to examine the consequences of the antiferromagnetic exchange. This led Kondo to calculate the magnetic scattering rate $\frac{1}{\tau}$ of electrons to cubic order in the *s-d* interaction. To his surprise, the cubic term contained a logarithmic temperature dependence[20]:

$$\frac{1}{\tau} \propto \left[J\rho + 2(J\rho)^2 \ln \frac{D}{T} \right]^2$$ (15)

where ρ is the density of states of electrons in the conduction sea and D is the half-width of the electron band. Kondo noted that if the *s-d* interaction were positive and antiferromagnetic, then as the temperature is lowered, the coupling constant, the scattering rate and resistivity start to rise. This meant that once the magnetic scattering overcame the phonon scattering, the resistance would develop a resistance minimum. Such resistance minima had been seen in metals for more than 30 years.[21,22] Through Kondo and Anderson's work, this 30 years old mystery could be directly interpreted as a a direct consequence of the predicted antiferromagnetic *s-d* interaction with local magnetic moments.

3.4. *Kondo's result poses a problem*

After Kondo and Anderson's work, the community quickly realized that the "Kondo effect" raised a major difficulty. You can see from (15) that at the "Kondo temperature" $T \sim T_K$ where $2J\rho \ln(D/T_K) \sim 1$, or

$$T_K \sim De^{-1/(2J\rho)}$$ (16)

the Kondo log becomes comparable with the bare interaction, so that at lower temperatures perturbation theory fails. What happens at lower temperatures once perturbation theory fails? This is the "Kondo problem".

By the late 1960s, from the work of early pioneers on the Kondo problem, including Alexei Abrikosov, Yosuke Nagaoka, Harry Suhl, Bob Schrieffer and Kei Yosida, much had been learned about the Kondo problem. It had become reasonably clear that at low temperatures the Kondo coupling constant grew to strong coupling, to form a spin singlet, but the community was divided over whether the residual scattering would be singular, or whether it would be analytic, forming an "Abrikosov Suhl" resonance. The problem also lacked lacked a conceptual framework and there were no controlled approximations.

3.5. *How a catastrophe led to new insight*

The solution to Kondo problem required a new concept — the renormalization group. Today we know the Kondo effect as an example of asymptotic freedom — a running coupling constant that flows from weak coupling at high energies, to strong coupling at low energies, ultimately binding the local moment into a singlet with electrons in the conduction sea. In the late 1960s, renormalization had entered condensed matter physics as a new tool for statistical physics. Phil and his collaborators now brought the renormalization group to quantum condensed matter by mapping the Kondo problem onto a one-dimensional Ising model with long range interactions.

Phil entered the field from an unexpected direction after discovering an effect known as the *orthogonality* or *X-ray* catastrophe. Phil's 1967 paper "Infrared Catastrophe in Fermi Gases with Local Scattering Potentials",[23] was stimulated by a conversation with John Hopfield, who speculated that the introduction of an impurity potential into a Fermi gas produces a new ground-state $|\phi^*\rangle$ that is orthogonal to the original ground-state $|\phi_0\rangle$. Phil examined this idea in detail, and showed that when a local scattering potential suddenly changes, in the thermodynamic limit, the overlap between the original and the new Fermi gas ground-states identically vanishes $\langle \phi_0 | \phi^* \rangle = 0$. For example, when an X-ray ionizes an atom in a metal, the ionic potential suddenly changes and this causes the conduction sea to evolve from its original ground-state $|\phi_0\rangle$ into a final-state $|\phi_f(t)\rangle = e^{-iHt}|\phi_0\rangle$.[24] In fact, Phil showed that the resulting relaxation is *critical*, with power-law decay in the overlap amplitude

$$G(t) = \langle \phi_0 | \phi_f(t) \rangle \sim \frac{e^{-i\Delta E_g t}}{t^\epsilon}, \tag{17}$$

where ΔE_g is difference between final and initial ground-state energies. The absence of a characteristic time scale indicates that the relaxation into the final-state ground-state is infinitely slow. By Fourier transform, this implies a singular density of states[24]

$$\rho(E) \sim \int dt G(t) e^{iEt} \sim (E - E_g)^{-1+\epsilon}. \tag{18}$$

This topic was also studied by Mahan[24] who linked the subject with X-ray line-shapes. Nozières and de Dominicis[25] later found an exact solution to the

integral equations of the orthogonality catastrophy. The X-ray catastrophe
is also responsible for the singular Green's functions of electrons in a
one-dimensional Luttinger Liquid.[26–28]

One of the key conclusions of this work was that the orthogonality catastrophe *occurs* in the Kondo problem. Phil recognized that each time a local
moment flips, the Weiss field it exerts on conduction electrons reverses, driving an orthogonality catastrophe in the "up" and "down" electron fluids. In
its anisotropic form, the Kondo interaction takes the form

$$H_K = J_z S_{dz} \sigma_z(0) + J_\perp \left[S_+ \sigma_-(0) + S_- \sigma_+(0) \right] \tag{19}$$

where $\sigma_\pm = (\sigma_x \pm i\sigma_y)/2$ and $S_\pm = S_{dx} \pm iS_{dy}$ are the local lowering and
raising operators for the mobile conduction and localized d-electrons respec-
tively. From the work of Nozières and de Dominicis, the amplitude for two
spin flips at times t_1 and t_2 is

$$(J_\perp)^2 \left(\frac{\tau_0}{t_2 - t_1} \right)^{2-2\epsilon} = (J_\perp)^2 \exp \left[-(2-\epsilon) \ln \left(\frac{t_2 - t_1}{\tau_0} \right) \right]. \tag{20}$$

where τ_0 is the short-time (ultra-violet) cut-off and $\epsilon \sim 2J_z\rho$ is determined
by the change in the scattering phase shift of the up and down Fermi gases,
each time the local spin reversed. This suggested that the quantum spin flips
in a Kondo problem could be mapped onto the statistical mechanics of a 1D
Coulomb gas of "kinks" with a logarithmic interaction.

3.6. *The Anderson–Yuval solution to the Kondo problem*

Working with graduate student Gideon Yuval[29,30] and a little later, Bell
Labs colleague Don Hamann, Phil's team took up the task of organizing and
summing the X-ray divergences of multiple spin-flip processes as a continuous
time path-integral. With some considerable creativity, it became possible to
map the quantum partition function of the Kondo model onto the *classical*
partition function of a Coulomb gas of kinks. By regarding the kinks as
domain walls in a one-dimensional Ising model, they could further map the
problem onto a one-dimensional Ising Ferromagnet spin chain with a $1/r^2$
interaction,

$$\frac{H}{T} = -(2-\epsilon) \sum_{i>j} \frac{S_i^z S_j^z}{|i - j|^2} - \mu \sum_i \sum_i S_i^z S_{i+1}^z, \tag{21}$$

where the Ising spins can have values $S_j = \pm 1/2$ at each site; the position
j along the chain is really the imaginary time $\tau = j\tau_0$ measured in units of
the short-time cut-off, with periodic boundary conditions and a total length

determined by the inverse temperature, $L = \frac{\hbar \tau_0}{k_B T}$. The tuning parameter $\epsilon = J_z \rho$ is determined by the Ising part of the exchange interaction, while the transverse interaction $J_\perp$ sets $\mu = -2 \ln J_\perp \rho$, the chemical potential of domain-wall kinks in the ferromagnetic spin chain. The larger $J_\perp \rho$, the more kinks are favored.

Suddenly a complex quantum problem became a tractable statistical mechanics model. It meant one could adapt the renormalization group from statistical physics to examine how the effective parameters of the Kondo parameter changed at lower and lower temperatures. By integrating out the effects of two closely separated pairs of spin flips, Anderson, Yuval and Hamann[31] derived the scaling equations

$$\frac{\partial J_z}{\partial \ln \tau_0} = J_\perp^2,$$

$$\frac{\partial J_\perp}{\partial \ln \tau_0} = J_\perp J_z. \tag{22}$$

Under these scaling laws, $J_\perp^2 - J_z^2$ is conserved, giving rise to the famous scaling trajectories shown in Fig. 3. There are two fixed points:

(a) $\epsilon \sim J_z \rho < 0$ Ferromagnetic ground-state $\equiv$ unscreened local moment.
(b) $\epsilon \sim J_z \rho > 0$ "Kink liquid" $\equiv$ screened local moment, where the Kondo temperature sets the typical kink separation $l_0 \sim T_K$.

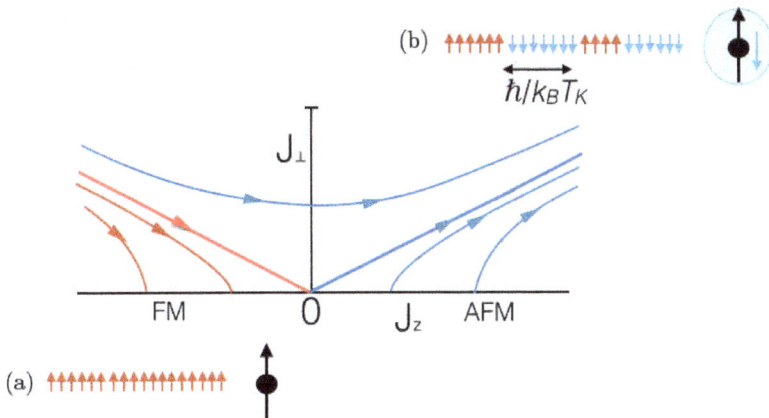

Fig. 3.   The Anderson–Yuval–Hamann scaling curves[31] for the Kondo model. (a) Red trajectories correspond to an unscreened moment, or a kink-free ferromagnetic ground-state in the magnetic analogy. (b) Blue trajectories correspond to a fully screened local moment, or a "kink condensate" where the average kink separation is determined by the inverse Kondo temperature.

In this way, the Coulomb gas of kinks had a phase transition at $J_z = 0$. For negative $J_z < 0$, the kinks are absent, but for positive J_z, the chemical potential of the kinks grows so that they proliferate, forming a kink-liquid. Although Anderson, Yuval and Hamann were unable to completely solve the strong coupling problem, the problem was solvable for so-called *Toulouse limit*, where $\epsilon = 2J_z\rho = 1$, and in this limit, it could be shown that the strong coupling limit was free of any singular scattering. In their paper,[31] the authors conclude that

> "The most interesting question on the Kondo effect has been from the start whether it did or did not fit into the structure of usual Fermi gas theory: In particular, does a true infrared singularity occur as in the X-ray problem, or does the Kondo impurity obey phase-space arguments as $T \to 0$ and give no energy dependences more singular than E^2 (or T^2), and is [the susceptibility] $\chi(T = 0)$ finite? The result we find is that the usual antiferromagnetic case in fact <u>does</u> fit after the time scale has been revised to τ_κ, i.e. that it behaves like a true bound-singlet as was conjectured originally by Nagaoka."

i.e the authors conclude that ground-state of the Kondo problem is a Fermi liquid.

During this period, Phil wrote series of informal papers in Comments in Solid State Physics[15,32–34] that provided a very personalized update on the progress. The last of these papers, "Kondo effect IV: out of the wilderness",[34] summarizes what become the *status quo* in this problem. Phil writes

> "In conclusion then, the status is this: we understand very clearly the physical nature of the Kondo problem, which is beautufully expressed in Fowler's picture of scaling: electrons of high enough energy interact with the weak, bare interaction and the bare Kondo spin, but as we lower the energy the effects of the other electrons gradually strengthen the effective interaction until finally, at energies near T_K, the effective interaction starts to get so large that we must allow the local spin to bind a compensating spin to itself, and the Kondo spin effectively disappears, being replaced by a large resonant non-magnetic scattering effect. My own opinion is that the low temperature behavior is totally non-singular, the Kondo impurity looking simply like a localized spin fluctuation site, but others believe that there may remain a trace of singular behavior."

The influence of these ideas ran far and wide:

(1) It introduced scaling theory to quantum systems. The project started by Anderson and Yuval later culminated in Wilson's numerical renormalization work.[35]

(2) The key conclusion about the non-singular character of the ground-state, confirmed confirmed by Wilson, later became the basis for Nozières' strong coupling treatment of the key low temperature properties of the Kondo problem,[37] which by mapping the physics onto

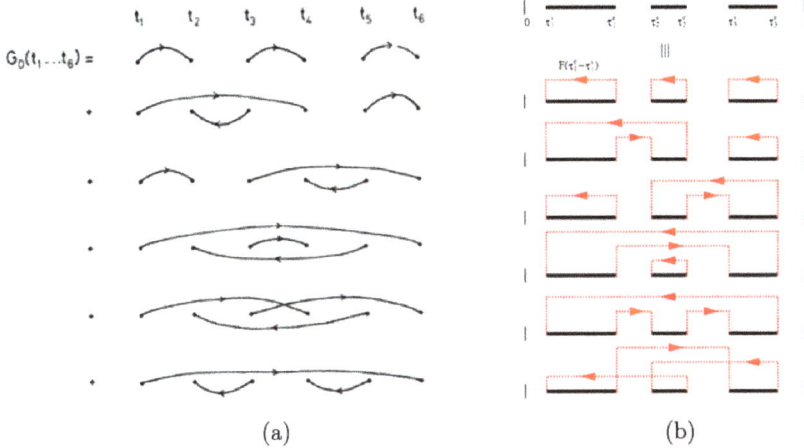

Fig. 4. Figure contrasting the use of the Yuval–Anderson approach to impurity models[29] with the modern continuous time Monte Carlo methods[36]: (a) illustrative sixth-order diagram from the original Anderson–Yuval paper,[29] (b) the same set of diagrams used in Ref. 36. The black-arrowed lines in (a) and the red-arrowed lines in (b) describe conduction electrons propagating between spin flips.

a local Fermi liquid, accounted for the two-fold enhancement of the Sommerfeld–Wilson ratio observed in Wilson's numerical work.[35]

(3) The statistical mechanics of the problem, with scale-dependent interactions between topological defects, provided inspiration and ground-work for Kosterlitz and Thouless's scaling solution[38] to the Berezinski–Kosterlitz–Thouless transition in the 2D xy antiferromagnet, whose scaling flows replicate the Anderson–Yuval–Hamann diagram.

(4) Phil's belief that Kondo model would be exactly solvable was dramatically confirmed by the independent Bethe Ansatz solutions of Natan Andrei and Paul Weigman[39,40]

(5) Modern continuous-time Montecarlo solvers for computational Dynamical Mean Field Theory approaches to materials research are a direct descendant of the Anderson–Yuval mapping of quantum impurity models to statistical mechanics in time (see Fig. 4).[36,41]

4. Green Period: Mixed Valence and the Large N Expansion

The solution to the Kondo problem resurrected the question about when fluctuations (quantum, random) defeat antiferromagnetic order, and when they do, what replaces it? Leaving the Kondo wilderness behind, Anderson's magnetic life developed in directions that explore this question. One way to

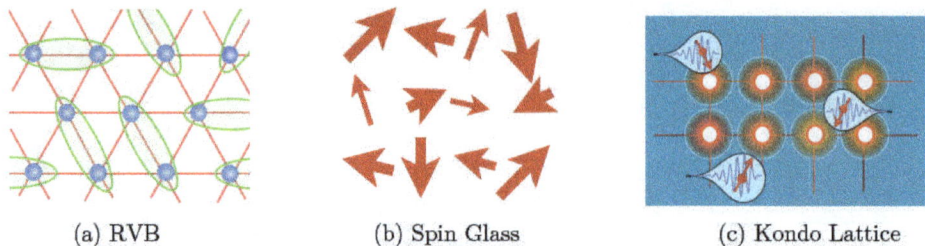

(a) RVB   (b) Spin Glass   (c) Kondo Lattice

Fig. 5. Three new directions for Anderson's research in the 1970s: (a) the resonating valence bond ground-state for the frustrated triangular lattice,[10] (b) the spin-glass ground-state for a frustrated disordered array of spins[43] and (c) the problem of mixed valence, where mobile heavy electrons move through a lattice of Kondo screened local moments.

avoid magnetism is by enhancing zero-point fluctuations with frustration, as in a triangular lattice antiferromagnet, and it was this direction Phil explored with Patrik Fazekas,[10] leading them to apply Pauling's resonating valence bond idea to spin liquids (see Fig. 5a). A second way is through quenched disorder, as in dilute magnetic alloys, and this led Phil, with Sam Edwards to invent the concept of the spin glass[42,43] (see Fig. 5b). A third direction, is through quantum fluctuations induced by the Kondo effect and valence fluctuations. This returned Anderson to the unfinished business of the Anderson and Kondo lattices (see Fig. 5c). The first two directions are discussed in the excellent articles by Ted Kirpatrick, Patrick Lee and Mohit Randeria in this volume. Here, I will focus the discussion on Phil's contributions to our understanding of the Kondo lattice and valence fluctuations.

In the 1970s experimentalists started to investigate the fate of dilute magnetic alloys as the magnetic atoms become more concentrated. In transition metal alloys, the RKKY interaction between the magnetic ions overcomes the Kondo effect, giving rise to spin glasses.[42,44] But in rare earth and actinide intermetallic compounds, the Kondo effect and associated valence fluctuations are strong enough to overcome the magnetism, even in fully concentrated lattices of local moments, leading to a wide variety of *heavy fermion* materials. Phil's insights played a vital role in the development of the field.

Early experimental progress in the new field of *mixed valence* was rapid and chaotic. A plethora of new intermetallic compounds were discovered which display local moment physics at high-temperatures, but which instead of magnetically ordering at low temperatures, form an alternative ground-state. Already in 1969, the group of Ted Geballe at Bell Labs had discovered SmB_6,[45] in which the magnetic Sm ions avoid ordering by developing a narrow-gap insulator, now called a "Kondo insulator". In 1975, an ETH Zurich-Bell Labs collaboration discovered the first heavy fermion metal $CeAl_3$.[46] The amazing thing about these two materials, is that both display the same sort of Kondo resistance scattering at high-temperatures, but at low temperatures two materials respond differently — with the resistivity sky-rocketing in SmB_6, but collapsing into a coherent low temperature metal in $CeAl_3$. Three years later, Frank Steglich discovered the first heavy fermion superconductor $CeCu_2Si_2$,[47] though it took a number of years for the community to change their mind-set and accept this pioneering discovery.

Despite this rapid progress on the experimental front, theoretical progress was flummoxed by the difficulty of making the transition to the dense "Kondo lattice" problem, lacking both the conceptual and mathematical framework. Phil's input, particularly his summary talks at the 1976 Rochester and 1980 Santa Barbara meetings on mixed valence had a profound impact.

Ron Parks and Chandra Varma organized the first conference on mixed valence at the University of Rochester in November 1976 and invited Phil to give the summary. As part of this summary, Phil roasted the theory community by sketching the resistivity of heavy fermion metals (see Fig. 6a) in the guise of an elephant. Recalling an Indian parable about an elephant and seven blind men, one who pulls its tail and says its a rope, the other who says its leg is tree and so on, Phil introduced his elephantine sketch of Kondo lattice resistivity, with Jun Kondo sliding down the elephant's trunk and a

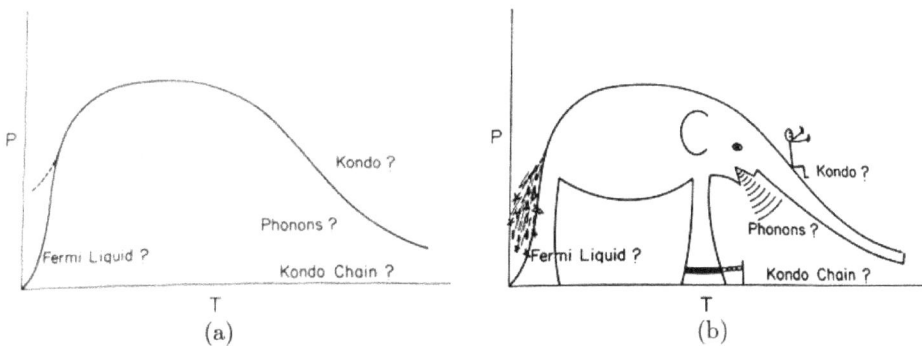

Fig. 6. Sketches from Phil Anderson's "Epilogue" from the 1976 Rochester Conference on Mixed Valence.[48] (a) Resistivity stereotypical of systems such as $CeAl_3$. (b) Elephantine version of (a). (Artwork by PWA). Reproduced from Valence Instabilities and Narrow Band Phenomenon, P. W. Anderson, ed. Ron Parks, pp. 389–396 (1977), with permission from the author.

Fermi liquid coming out of its behind! (See Fig. 6b).[b] The main point of the figure however, was to urge the theory community to unify its understanding of these diverse phenomena.

In the transcript, describing the Kondo elephant, Phil remarks

"Now we come to the heart and core of the elephant, the part which nobody has really done, which was first mentioned at least as a serious problem here in this conference, namely the Kondo lattice, which Seb Doniach has made a start on. What you really have here is a lattice full of these objects that fluctuate back and forth from one valence to another. There are the phonons, there is the fact that the electrons fluctuate by tossing electrons into the d level on the next site which can then go down into the f levels on yet another site. So the things which toss the valence back and forth are definitely coupled between one site and another. The net result of doing this is something that most of the experiments have to tell us about: that this probably renormalizes to a very heavy Fermi liquid theory with some kind of strong antiferromagnetic prejudice in that the f-like objects in the Fermi liquid somehow lost all of their desire to be magnetic and don't very easily order anymore. This is an extremely hard problem, it's a problem in the same category of problems which are failing to be done in field theory these days."

In this brief paragraph, Phil has laid out his view of the physical framework needed to understand heavy fermion materials. Despite his cartoon, he did emphasize that the low temperature ground-state would be a renormalized heavy Fermi liquid. He also notes the parallel between strongly correlated materials and the challenges of field theory, a parallel that would inspire many younger physicists in the decades to come. Later in the talk, Phil discusses the possibility of further instabilities in the Fermi liquid, and expresses the view that these will be more than just antiferromagnets:

"Once you get down to this Fermi liquid, it seems that there is a serious question of what then happens? What does the resulting heavy Fermi liquid do with itself, what further transformation might it undergo? There are several possibilities. . . . There is no reason at all why it shouldn't localize and maybe there are cases where it localizes. Kasuya gave an argument for one of then. A second possibility a whole series of experiments seem to indicate is that some phase transition takes place in many cases. The question is: what is the nature of these phase transitions? I for one am not ready to accept the idea that they are all simple magnetic phase transitions . . . . Maybe there is some kind of d to f excitonic phase transition that either does or does not leave some Fermi surface behind. Maybe there's a density wave. What else?"

---

[b]Did this sketch reveal Phil's subconscious discomfort with Landau Fermi liquid theory?

Curiously though, reflecting the continued mind-set of the community Phil does not mention the possibility of superconductivity, reflecting the fact that Steglich's 1979 work was not yet widely accepted.[c]

In 1980, Walter Kohn, Brian Maple and Werner Hanke at the Institute for Theoretical Physics, Santa Barbara (now the Kavli Institute for Theoretical Physics) organized a six month workshop on valence fluctuations, culminating in a conference in January 1981. In the summary of the conference, Phil continued on the theme of the link between field theory and strongly correlated electrons, introducing for the first time, the seminal idea that a *large N expansion*, akin to that used in particle physics, might be useful.

Phil magnetic life already had two links with the idea of a large N expansion. Of course, his early work on spin-wave theory was based on a $1/S$ expansion, but more recently, his work with Sam Edwards on the infinite range spin glass had involved the replica trick, replacing the disorder-averaged Free energy $\overline{-T \ln Z}$ with the $N \to 0$ limit of the disorder-averaged partition function average of N replicas,

$$\overline{\ln Z} = \lim_{N \to 0} \overline{(Z^N - 1)}. \tag{23}$$

To take the $N \to 0$ limit requires that one first solve the problem at large N to extrapolate back to zero.

But the context of heavy fermions Phil noticed that there was already a large finite N to expand in. Rare earth atoms are strongly spin-orbit coupled, and so, ignoring crystal field effects, they have a large spin degeneracy $N = 2j + 1$, where $j = 5/2$ or $7/2$ for individual f-electrons. Phil realized that the parameter $1/N$ could act as an effective small parameter for resuming many body effects:

> "The most important one, ... is the importance of what you might call the large N limit; it was only at this conference that I, at least, realized that we have been going through a case of parallel evolution with non-Abelian gauge theory. This really has great resemblances to what one does in the intermediate valence problem, and it is interesting that the gauge theorists have found that their best controlled approximations are in a limit which they call large N — which is large order of the group, large degeneracy of the particles, and in our case that has to do with large values of the degeneracies of the states. This is the number that Ramakrishnan called n_λ. I'm going to talk later about how many different kinds of roles that plays."

---

[c]Indeed, although Phil didn't know it, superconductivity had been seen at Bell Labs three years earlier in the heavy fermion material UBe_{13},[49] but mis-interpreted as an artifact of uranium filaments.

Later in the same article, Phil expands on this idea and how it can be used for scaling. He makes two key observations:

- That valence fluctuations are N-fold enhanced by the large orbital degeneracy of f-electrons.
- That intersite interactions are reduced by a factor of $1/N$ relative to onsite interactions.

Summarizing a full page of discussion, Phil writes

"So we find again and again that we are gaining from this degeneracy factor and it may make the problem a lot simpler than such apparently easier problems like the Kondo problem."

Phil's new proposal had an electrifying effect on the fledgling strongly correlated electron theory community, for it undid the log-jam, providing for the first time, a controllable expansion parameter for dealing with the mixed valent and Kondo lattices. In the immediate future, a wide range of large N treatments of the Anderson model followed, including work by Ramakrishnan and Sur,[50,51] Gunnarson and Schonhammer[52] and by Zhang and Lee.[53] Phil's observations also inspired a search for a more field theoretic way to formulate the Kondo and mixed valence problems, leading to the pioneering work by Nicolas Read and Dennis Newns[54,55] on the large N Kondo model and my own *slave boson* approach[56] to mixed valence developed under Phil's generous tutelage, in which the Gutzwiller projected f-electron operator is factorized in terms of an Abrikosov pseudo-fermion and a slave boson operator $X_{\sigma 0} = f_\sigma^\dagger b$. With this device, one could see for the first time, that the no-double occupancy constraint gives rise to locally conserved charges (here $n_b + n_f = Q$) and corresponding gauge fields.

5. Superconductivity and Magnetism Come Together

The theoretical perspective of condensed matter physics has dramatically transformed over the period of Phil's research. Fifty years ago, magnetism and superconductivity were regarded as mutually exclusive forms of order. Yet, gradually, starting in the 1970s, the discovery of new kinds of pair condensate, of superfluid ^{3}He,[58] of heavy fermion,[47] organic,[59] high-temperature cuprate[60] and iron-based superconductors,[61] has indicated a more intimate connection between magnetism and superconductivity.

Phil's ideas have evolved during this same period, and as they have done so, they have often transformed our scientific consensus. Phil started in the 1950s accounting for the stability of antferromagnetic order against quantum fluctuations, at the time itself contraversial. Through a journey via the Kondo effect, spin liquids and spin glasses he was led to consider states of matter in which conventional magnetism is absent and the magnetic degrees of freedom drive new kinds of electronic ground-states, especially superconductivity.

I'd like to selectively mention three exciting areas of evolving and currently unresolved controversy connected with Phil's ideas.

(1) Mott versus Landau. From the very outset, Phil he has emphasized the importance of the *Mott mechanism*, namely the exclusion of double occupancy of atomic orbitals. One of the questions he has emphasized, is whether the Landau quasiparticle description of electrons can survive the imposition of these severe constraints, suggesting instead that new kinds of metallic ground-states must inevitably develop in which the excitations have zero overlap with non-interacting electrons, and thus can not be regarded as Landau quasiparticles. Central to Phil's arguments, is the idea that the of electrons to highly constrained electron fluids leads to a many-body X-ray catastrophe that leads to the innevitable demise of the Landau quasiparticle to form *strange metals*,[62,63] and sometimes, *hidden Fermi liquids*,[64,65] which resemble the Landau Fermi liquid thermodynamically, but without overlap with the original electron fields.

(2) Quantum Criticality versus Strange Metals. There are now many example of metals which exhibit highly unusual transport and thermodynamic properties which defy a Landau Fermi liquid description, such as the optimally doped normal state of the cuprate superconductors,[66] MnSi under pressure and various heavy fermion materials,[67] such as $CeCoIn_5$[57,68] (see Fig. 7) and $YbAlB_4$[69] which each exhibit unusual linear or power-law temperature dependencies in the resitivity. One of the key discussions about these materials is whether such non-Fermi liquid behavior is generated by the vicinity to a *quantum critical point*, or whether, as Phil believes, the unusual metallic behavior is related to a new kind *strange metal phase*.[65] The recent discovery of a pressure-independent anomalous metal phase in $YbAlB_4$ may be an example of such a strange metal phase.[69]

(3) Fabric versus Glue. The conventional view of unconventional superconductors argues that they should be regarded as magnetic analogs of phonon-mediated superconductors, in which the soft magnetic fluctuations provide the pairing *glue*. Phil has argued[70] for a different picture, in which pre-formed, resonating valence bonds, on doping,

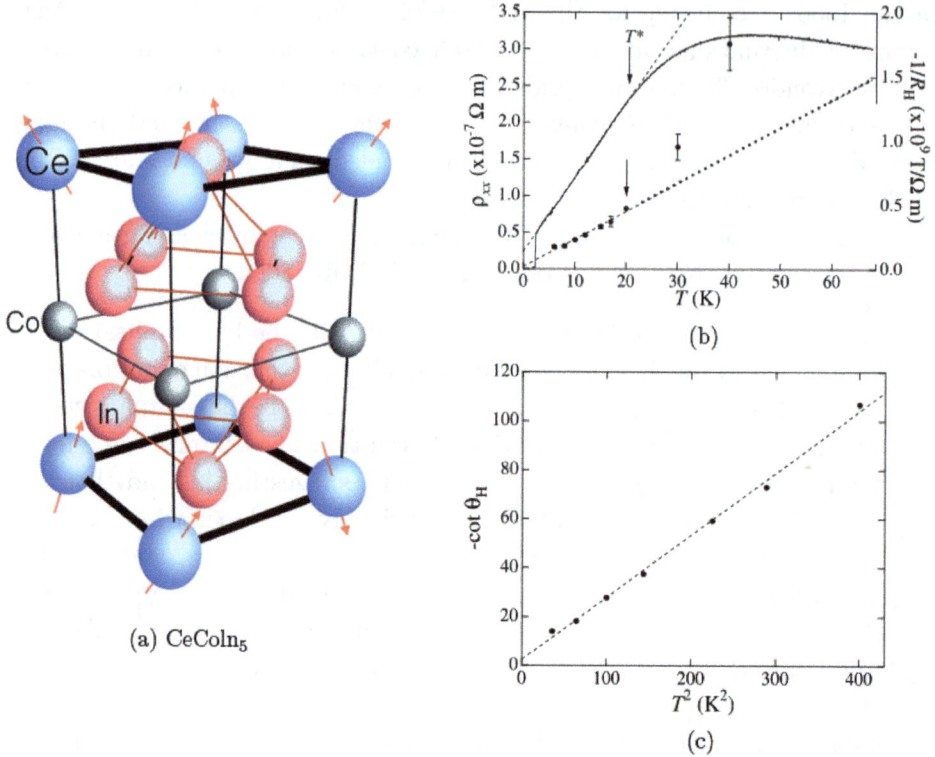

(a) CeCoIn$_5$

(b)

(c)

Fig. 7. The heavy fermion superconductor CeCoIn$_5$ showing (a) the structure of this layered compound (b) the resistivity and inverse Hall constant, which are both linearly proportional to the temperature. The linear temperature dependence of the resistivity indicates a linear temperature dependence of the electron scattering rate. The temperature dependence of the Hall constant indicates that the Hall transport relaxation rate and the linear transport relaxataion rate are not equal after.[57] (c) the cotangent of the Hall angle, showing the T^2 dependence of the Hall scattering rate after.[57]

provide the underlying *fabric* for a pair condensate. These opposing ideas continue to be lively debated in the context of high-temperature cuprate superconductors. Another place they may be important, is in heavy fermion superconductors, where the Kondo effect can play the same role as doping, forcing valence bonds out into the conduction sea to form pairs.

In his 2006 paper, "The strange metal is a projected Fermi liquid with edge singularities",[63] Phil summarizes his point of view, writing

"This strange metal phase continues to be of much theoretical interest. Here we show it is a consequence of projecting the doubly occupied amplitudes out of a conventional Fermi-sea wave function (Gutzwiller projection), requiring no exotica such as a mysterious quantum critical point. Exploiting a formal similarity with the classic problem of Fermi-edge singularities in the X-ray spectra of metals, we find a Fermi-liquid-like excitation spectrum, but the excitations are asymmetric between electrons and holes, show anomalous forward scattering and the renormalization constant $Z = 0$."

One of the most fascinating, and still unsolved aspects of these discussions above concerns the apparent development of two transport lifetimes in the electronic conductivity[66,71]: a transport scattering lifetime, inversely proportional to temperature $\tau_{\rm tr}^{-1} \propto k_B T$ and a Hall scattering time, inversely proportional the square of the temperature $\tau_H^{-1} \propto T^2$. In a modified Drude formalism, the linear and Hall conductivities are given by

$$\sigma_{xx} = \frac{ne^2}{m}\tau_{\rm tr}$$

$$\sigma_{xy} = \frac{ne^2}{m}\tau_{\rm tr}(\omega_c \tau_H), \tag{24}$$

giving rise to a resistivity $\rho_{xx} \propto \tau_{\rm tr}^{-1} \sim T$ and a Hall angle which satifies $\cot\theta_H = \sigma_{xx}/\sigma_{xy} \propto \tau_H^{-1} \propto T^2$. There are now three separate classes of material where this behavior has been seen: the cuprate metals,[66] the 115 heavy fermion superconductors CeCoIn$_5$[57] and electrons fluids at two-dimensional oxide interfaces (SrTiO$_3$/RTiO$_3$ (R = Gd, Sm)).[72] The remarkable aspect of these metals, is that the two relaxation times enter *multiplicatively* into their Hall conductivity, $\sigma_{xy} \propto \tau_{\rm tr}\tau_H$. Since since σ_{xy} is a zero momentum probe of the current fluctuations at the Fermi surface, this suggests that electrons are subject to two separate relaxation times at the very same point on the Fermi surface, linked by the current operator. Phil's ideas on this subject[71] have inspired a range of new theories,[73-75] but we still await a final understanding.

Like many in our community, I've often marvelled at Phil Anderson's ability to radically transform his viewpoints in response to new data and new insights. I've asked him what it would be like if he ever met his younger self for a physics discussion, and he agrees that he'd probably have quite a forceful disagreement on topics he originally pioneered and on which he now has a new perspective. Perhaps Tom Stoppard will write a play on this someday.

Phil, here's to the continuing success and inspiration of your magnetic ideas!

Acknowledgments

In writing this article I have benefited from conversations with Natan Andrei, Premala Chandra and Don Hamann. This work was supported by NSF grant DMR-1309929.

References

1. P.W. Anderson, *Basic Notions of Condensed Matter Physics* (Benjamin Cummings, 1984).
2. L. Néel, Influence des fluctuations du champ molculaire sur les proprits magntiques des corps, *Ann. de Physique* **18**, 5 (1932).
3. L.D. Landau, A possible explanation of the field dependence of the suseptibility at low temperatures, *Phys. Z. Sowjet.* **4**, 675 (1933).
4. H. Bethe, Zur Theorie der Metalle: I. Eigenwerte und Eigenfunktionen der linearer Atomkerte (On the theory of metals: I. Eigenvalues and Eigenfunctions of the linear atom chain), *Zeitschrift fur Physik.* **71**, 205–226 (1931).
5. I. Pomeranchuk, Thermal conductivity of paramagnetic insulators at low temperatures, *Zh. Eksp. Teor. Fiz.* **11**, 226 (1941).
6. P.W. Anderson, An Approximate Quantum Theory of the Antiferromagnetic Ground State, *Phys. Rev.* **86**, 694–701 (1952).
7. C.G. Shull and J.S. Smart, Detection of Antiferromagnetism by Neutron Diffraction, *Phys. Rev.* **76**, 1256–1257 (1949).
8. M.J. Klein and R. Smith, A Note on the Classical Spin-Wave Theory of Heller and Kramers, *Phys. Rev.* **80**, 1111 (1950).
9. R. Kubo, The Spin-Wave Theory of Antiferromagnetics, *Phys. Rev.* **87**, 568–580 (1952).
10. P. Fazekas and P.W. Anderson, On the ground state properties of the anisotropic triangular antiferromagnet, *Philos. Mag.* **30**, 423–440 (1974).
11. F.D.M. Haldane, Continuum dynamics of the 1-D Heisenberg antiferromagnet: Identification with the O(3) nonlinear sigma model, *Phys. Lett. A* **93**, 464–468 (1982).
12. P.W. Anderson, Random-Phase Approximation in the Theory of Superconductivity, *Phys. Rev.* **112**, 1900–1916 (1958).
13. Y. Nambu, Quasi-Particles and Gauge Invariance in the Theory of Superconductivity, *Phys. Rev.* **117**, 648–663 (1960).
14. P.W. Anderson, New Approach to the Theory of Superexchange Interactions, *Phys. Rev.* **115**, 2–13 (1959).
15. P.W. Anderson, The Kondo Effect II, *Comments on Solid State Physics.* **I**, 190 (1968–1969).
16. P.W. Anderson. The Kondo Effect II, in *A Career in Theoretical Physics* (World Scientific, 1994), p. 224.
17. C. Zener, Interaction between the *d* Shells in the Transition Metals, *Phys. Rev.* **81**, 440–444 (1951).
18. T. Kasuya, A Theory of Metallic Ferro- and Antiferromagnetism on Zener's Model, *Prog. Theor. Phys.* **16**(1), 45–57 (1956).

19. J.R. Schrieffer and P. Wolff, Relation between the Anderson and Kondo Hamiltonians, *Phys. Rev.* **149**, 491 (1966).

20. J. Kondo, Resistance Minimum in Dilute Magnetic Alloys, *Prog. Theor. Phys.* **32**, 37–49 (1964).

21. de Haas, de Boer and D. van den Berg, The electrical resistance of gold, copper and lead at low temperatures, *Physica* **1**, 1115 (1933).

22. D. MacDonald and K. Mendelssohn, Resistivity of Pure Metals at Low Temperatures I. The Alkali Metals, *Proc. Roy. Soc. London.* **202**, 523 (1950).

23. P.W. Anderson, Infrared Catastrophe in Fermi Gases with Local Scattering Potentials, *Phys. Rev. Lett.* **18**, 1049–1051 (1967).

24. G. Mahan, Excitons in Metals: Infinite Hole Mass, *Phys. Rev.* **163**(3), 612–617 (1967).

25. P. Nozières and C.T. de Dominicis, Singularities in the X-Ray Absorption and Emission of Metals. III. One-Body Theory Exact Solution, *Phys. Rev.* **178**, 1097–1107 (1969).

26. P. Anderson, Infrared Catastrophe: When Does It Trash Fermi Liquid Theory? In D. Baeriswyl, D. Campbell, J. Carmelo, F. Guinea and E. Louis (Eds.), *The Hubbard Model*, Vol. 343, NATO ASI Series (Springer, 1995), pp. 217–225.

27. A. Imambekov and L.I. Glazman, Universal Theory of Nonlinear Luttinger Liquids, *Science* **323**(5911), 228–231 (2009).

28. G.A. Fiete, Singular responses of spin-incoherent Luttinger liquids, *J. Phys.: Condens. Matter* **21**(19), 193201 (2009).

29. P.W. Anderson and G. Yuval, Exact Results in the Kondo Problem: Equivalence to a Classical One-Dimensional Coulomb Gas, *Phys. Rev. Lett.* **45**, 370 (1969).

30. P.W. Anderson and G. Yuval, Exact Results for the Kondo Problem: One-Body Theory and Extension to Finite Temperature, *Phys. Rev. B* **1**, 1522 (1970).

31. P.W. Anderson, G. Yuval and D. Hamann, Exact Results in the Kondo Problem. II. Scaling Theory, Qualitatively Correct Solution, and Some New Results on One-Dimensional Classical Statistical Models, *Phys. Rev. B* **1**(11), 4464–4473 (1970).

32. P.W. Anderson, The Kondo Effect. I, *Comments on Solid State Physics.* **I**, 31 (1968–1969).

33. P.W. Anderson, The Kondo Effect III: The Wilderness — Mainly Theoretical, *Comments on Solid State Physics* **3**, 153 (1971).

34. P.W. Anderson, Kondo Effect IV: Out of the Wilderness, *Comments on Solid State Physics* **5**, 73 (1973).

35. K.G. Wilson, The renormalization group: Critical phenomena and the Kondo problem, *Rev. Mod. Phys.* **47**, 773 (1975).

36. P. Werner, A. Comanac, L. de' Medici, M. Troyer and A.J. Millis, Continuous-time solver for quantum impurity models, *Phys. Rev. Lett.* **97**, 076405 (2006). doi:10.1103/PhysRevLett.97.076405.

37. P. Nozières, A "Fermi Liquid" Description of the Kondo Problem at Low Tempertures, *Journal de Physique C* **37**, 1 (1976).

38. J.M. Kosterlitz and D.J. Thouless, Ordering, metastability and phase transitions in two-dimensional systems, *J. Phys. C: Solid State Physics* **6**(7), 1181 (1973).

39. N. Andrei, Diagonalization of the Kondo Hamiltonian, *Phys. Rev. Lett.* **45**, 379–382 (1980).

40. P.B. Weigman, Exact solution of sd exchange model at $T = 0$, *JETP Lett.* (1980).

41. K. Haule, Quantum Monte Carlo impurity solver for cluster dynamical mean-field theory and electronic structure calculations with adjustable cluster base, *Phys. Rev. B* **75**, 155113 (2007).

42. P.W. Anderson, Localisation theory and the CuMn problem: Spin glasses, *Mater. Res. Bull.* **5**, 549 (1970).

43. S.F. Edwards and P.W. Anderson, Theory of spin glasses, *J. Phys F: Metal Phys.* **6**, 965 (1975).

44. V. Cannella and J.A. Mydosh, Magnetic Ordering in Gold-Iron Alloys, *Phys. Rev. B* **6**(11), 4220–4237 (1972).

45. A. Menth, E. Buehler and T.H. Geballe, Magnetic and Semiconducting Properties of SmB_6, *Phys. Rev. Lett.* **22**, 295 (1969).

46. K. Andres, J. Graebner and H.R. Ott, 4f-Virtual-Bound-State Formation in CeAl$_3$ at Low Temperatures, *Phys. Rev. Lett.* **35**, 1779 (1975).

47. F. Steglich, J. Aarts, C.D. Bredl, W. Leike, D.E.M.W. Franz and H. Schäfer, Superconductivity in the Presence of Strong Pauli Paramagnetism: CeCu$_2$Si$_2$, *Phys. Rev. Lett.* **43**, 1892 (1979).

48. P.W. Anderson. *Epilogue*, in R.D. Parks (Ed.), *Valence Instabilities and Narrow-Band Phenomena* (Plenum, NY, 1977), pp. 389–396.

49. E. Bucher, J.P. Maita, G.W. Hull, R.C. Fulton and A.S. Cooper, Electronic properties of beryllides of the rare earth and some actinides, *Phys. Rev. B* **11**, 440 (1975).

50. T.V. Ramakrishnan, in L. M. Falicov, W. Hanke and M. P. Maple (Eds.), *Valence Fluctuations in Solids* (North Holland, 1981), p. 13.

51. T.V. Ramakrishnan and K. Sur, Theory of a mixed-valent impurity, *Phys. Rev. B.* **26**, 1798–1811 (1982).

52. O. Gunnarsson and K. Schönhammer, Electron spectroscopies for Ce compounds in the impurity model, *Phys. Rev. B* **28**, 4315 (1983).

53. F.C. Zhang and T.K. Lee, $\frac{1}{N}$ expansion for the degenerate Anderson model in the mixed-valence regime, *Phys. Rev. B* **28**, 33–38 (1983).

54. N. Read and D. Newns, On the solution of the Coqblin-Schrieffer Hamiltonian by the large-N expansion technique, *J. Phys. C.* **16**, 3274 (1983).

55. N. Read and D.M. Newns, A new functional integral formalism for the degenerate Anderson model, *J. Phys. C* **29**, L1055 (1983).

56. P. Coleman, 1/N expansion for the Kondo lattice, *Phys. Rev.* **28**, 5255 (1983).

57. Y. Nakajima, K. Izawa, Y. Matsuda, S. Uji, T. Terashima, H. Shishido, R. Settai, Y. Onuki and H. Kontani, Normal-state Hall Angle and Magnetoresistance in quasi-2D Heavy Fermion CeCoIn$_5$ near a Quantum Critical Point, *J. Phys. Soc. Japan* **73**, 5 (2004).

58. D.D. Osheroff, R.C. Richardson and D.M. Lee, Evidence for a New Phase of Solid He^3, *Phys. Rev. Lett.* **28**, 885–888 (1972).

59. D. Jrome, M. Mazaud, A. Ribault, and K. Bechgaard, Superconductivity in a synthetic organic conductor: (TMTSF)$_2$PF$_6$, *J. Phys. Lett. (Paris)* **41**, L95 (1980).

60. J.G. Bednorz and K.A. Muller, Possible high-T_c superconductivity in the Ba-La-Cu-O system, *Z. Phys.* **B64**, 189–193 (1986).

61. Y. Kamihara, H. Hiramatsu, M. Hirano, R. Kawamura, H. Yanagi, T. Kamiya and H. Hosono, Iron-Based Layered Superconductor: LaOFeP, *J. Am. Chem. Soc.* **128**(31), 10012–10013 (2006).

62. P.W. Anderson, "Luttinger-liquid" behavior of the normal metallic state of the 2D Hubbard model, *Phys. Rev. Lett.* **64**(15), 1839–1841 (1990).

63. P.W. Anderson, The "strange metal" is a projected Fermi liquid with edge singularities, *Nat. Phys.* **2**(9), 626–630 (2006).

64. P.W. Anderson, Hidden Fermi liquid: The secret of high-T_c cuprates, *Phys. Rev. B* **78**(17), 174505 (2008).

65. P.W. Anderson, Fermi Sea of Heavy Electrons (a Kondo Lattice) Is Never a Fermi Liquid, *Phys. Rev. Lett.* **104**(17), 176403 (2010).

66. T.R. Chien, Z.Z. Wang and N.P. Ong, Effect of Zn impurities on the normal-state Hall angle in single-crystal $YBa_2Cu_{3-x}Zn_xO_{7-\delta}$, *Phys. Rev. Lett.* **67**, 2088 (1991).

67. C. Pfleiderer, S.R. Julian and G.G. Lonzarich, Non-Fermi-liquid nature of the normal state of itinerant-electron ferromagnets, *Nature* **414**(6862), 427–430 (2001).

68. M.A. Tanatar, J. Paglione, C. Petrovic and L. Taillefer, Anisotropic Violation of the Wiedemann-Franz Law at a Quantum Critical Point, *Science* **316**(5829), 1320–1322 (2007).

69. Y. Matsumoto, S. Nakatsuji, K. Kuga, Y. Karaki and N. Horie, Quantum Criticality Without Tuning in the Mixed Valence Compound β-YbAlB4, *Science* (2011).

70. P.W. Anderson, *Physics.* Is there glue in cuprate superconductors? *Science* **316**(5832), 1705–1707 (2007).

71. P.W. Anderson, Hall effect in the two-dimensional Luttinger liquid, *Phys. Rev. Lett.* **67**, 2092 (1991).

72. E. Mikheev, C.R. Freeze, B.J. Isaac, T.A. Cain and S. Stemmer, Separation of transport lifetimes in SrTiO3-based two-dimensional electron liquids, *Phys. Rev. B* **91**(16), 165125 (2015).

73. P. Coleman, A.J. Schofield and A.M. Tsvelik, Phenomenological transport equation for the cuprate metals, *Phys. Rev. Lett.* **76**, 1324 (1996).

74. N.E. Hussey, Phenomenology of the normal state in-plane transport properties of high-T_c cuprates, *J. Phys.: Condens. Matter* (2008).

75. M. Blake and A. Donos, Quantum Critical Transport and the Hall Angle in Holographic Models, *Phys. Rev. Lett.* **114**(2), 021601 (2015).

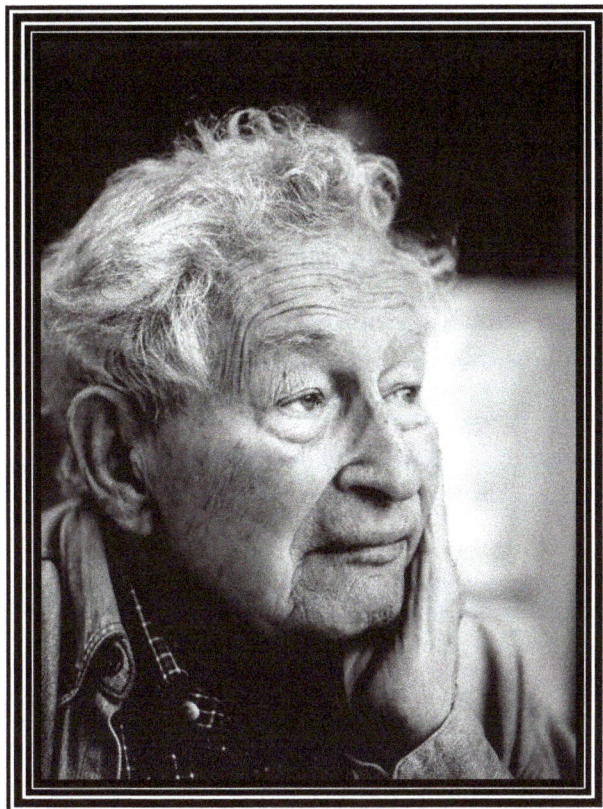

PWA90 Workshop

Marking the scientific accomplishments of
Philip W. Anderson
December 14-15, 2013
Princeton University

Taylor Auditorium, Frick Chemistry Laboratory
Sponsors:
Gordon and Betty Moore Foundation
National Science Foundation
Princeton University Department of Physics
Princeton Center for Complex Materials

Conference Program

Philip W. Anderson
Professor of Physics, winner of the 1977 Nobel Prize in Physics

In 1949, Philip W. Anderson published his first paper "Pressure Broadening in the Microwave and Infra-Red Regions" which discussed exchange and motional narrowing. His latest paper (No. 508) was uploaded a few weeks ago. In a remarkable career spanning more than six decades, Anderson has spawned, shaped, or made fundamental contributions to many subfields of condensed-matter physics. These range from Anderson localization, superconductivity, superfluidity, magnetism, ferroelectrics, to the physics of cuprates, spin glasses, and heavy-fermions. Many of the deep notions Anderson introduced or brought to fruition – dirty superconductivity, scaling ideas in localization and the Kondo problem, local moments, pseudospin, resonating valence bonds, broken symmetry and rigidity, spin liquid, frustration, superexchange, and orthogonality catastrophe — are now part of the standard lexicon of the field. Several generations of physicists have been strongly influenced by these concepts. Anderson's ideas have also extended deeply into other areas of physics (Anderson Higgs mechanism and the dynamics of pulsars). As codified in his oft-quoted phrase "More is Different," Anderson has been the most forceful and persuasive proponent of the radical (in 1970), but now ubiquitous, viewpoint of emergent phenomena: truly fundamental concepts can and do emerge from investigations of Nature at each layer of complexity or energy scale.

Co-Chairs:
Piers Coleman (coleman@physics.rutgers.edu)
N. Phuan Ong (npo@princeton.edu)
Committee Members:
Ravin Bhatt (ravin@princeton.edu)
Premi Chandra (pchandra@physics.rutgers.edu)
Duncan Haldane (haldane@princeton.edu)
David Huse (huse@princeton.edu)
Shivaji Sondhi (sondhi@princeton.edu)
Ali Yazdani (yazdani@princeton.edu)

8:00   Continental Breakfast, Frick Chemistry Lab, Atrium
9:00   Opening Remarks: Piers Coleman and N. Phuan Ong

Anderson Localization
Session Chair - Bhatt, Ravindra
9:10   "Some Recollections on Localization"
        Elihu Abrahams, UCLA
9:50   "Anderson Localization as a central paradigm in Quantum Physics"
        Boris Altshuler, Columbia University
10:30   Break

The ABEGHHK'tH Mechanism
Session Chair – William Brinkman
10:45   "Phil Anderson's Work On Gauge Symmetry Breaking"
        Ed Witten, Inst. Advanced Studies, Princeton
11:25   "Superfluidity and Symmetry Breaking: An Anderson Living Legacy"
        Frank Wilczek, MIT
12:05   Lunch, Frick Chemistry Lab, Atrium

Frustration, Magnetism, Heavy Fermions
Session Chair – David Huse
1:30   "40 years of quantum spin liquid: a tale of emergence from frustration"
        Patrick Lee, MIT
2:10   "Frustration and Magnetism"
        Scott Kirkpatrick, The Hebrew University of Jerusalem
2:50   "Anderson under the Microscope"
        Ali Yazdani, Princeton University
3:30   Break

Panel Discussion: Physicists on Science Policy and the world beyond
Session Chair – Piers Coleman
3:45   "Four Years in Washington"
        Bill Brinkman, Princeton University
        "Recollections of a young Phil Anderson"
        Walter Kohn, The University of California, Santa Barbara
        "Program management from the perspective of a government scientist"
        Pierre Morel, University of Paris
        "Modeling risk"
        Joe Wheatley, Managing Director, Biospherica Risk Ltd
        "Life outside physics on the dark side and beyond - from derivatives
        trading to angel investing"
        Zou Zhou, Hedge Fund Manager at Taconic Capital
5:30   Reception & Dinner, Prospect House - *Pre-registration required -
        CLOSED*

Sunday, December 15, 2013

8:00   Continental Breakfast, Frick Chemistry Lab, Atrium

Strong correlation, Superconductivity
Session Chair – Ali Yazdani

9:00   "Title TBA"
       T. Maurice Rice, Eidgenössische Technische Hochschule
9:40   "Theory of bad metallic behavior in oxides and pnictides"
       Gabi Kotliar, Rutgers University
10:20   Break

10:35   "What does RVB have to do with cuprate superconductors?"
       Juan Carlos Campuzano, University of Illinois at Chicago
11:15   "High Tc Superconductivity and RVB"
       Mohit Randeria, Ohio State University
11:55   Lunch, Frick Chemistry Lab, Atrium

Superfluids, Entanglement, Biophysics
Session Chair – Clare Yu
1:30   "From bacteria to artificial cells, the problem of self reproduction"
       Albert J. Libchaber, Rockefeller University
2:10   "The structure of a world (which may be) described by quantum
       mechanics"
       Anthony Leggett, Univ. of Illinois, Urbana-Champaign
2:50   "Geometry, topology, and Mott-Hubbard incompressibility in
       the Fractional Quantum Hall fluid."
       Duncan Haldane, Princeton University
3:30   Break
3:45   "Scenarios for Superconducting Protons in Pulsars"
       Mal Ruderman, Columbia University

Panel Discussion: Condensed Matter Physics: Historical & Grand Challenges
Session Chair - Premi Chandra
4:30   "How scientists can have a stronger voice in politics - my experiences
       on the way from physics to parliament"
       Ted Hsu, Member of Parliament, Canada
       "Anderson and Condensed Matter Physics"
       TV Ramakrishnan, Benaras Hindu University
       "PWA's mentoring at Cambridge, at ICTP Trieste, and everywhere
       else"
       Erio Tosatti, SISSA
       "Anderson localization, déjà vu all over again"
       David Thouless, University of Washington
6:00   Casual Dinner Reception, Frick Chemistry Lab, Atrium (No pre-
       registration required)

List of Conference Participants

Name	Institution
Abrahams, Elihu	UCLA
Aleksandrov, Victor	Rutgers University
Allen, Philip	Stony Brook University
Alpar, Ali	Sabanci University
Altshuler, Boris	Columbia University
Anderson, Phil	Princeton University
Andrei, Eva	Rutgers University
Andrei, Natan	Rutgers University
Artyukhin, Sergey	Rutgers University
Austin, Robert	Princeton University
Baskaran, Ganapathy	Institute of Mathematical Sciences, India
Bedell, Kevin	Boston College
Belopolski, Ilya	Princeton University
Ben-Benjamin, Jonathan	CUNY
Bernevig, B. Andrei	Princeton University
Bezdetko, Ilya	Stevens Institute of Technology
Bhatt, Ravindra	Princeton University
Bialek, William	Princeton University
Bian, Guang	Princeton University
Blagoev, Krastan	National Science Foundation
Brinkman, William	Princeton University
Calan, Curtis	Princeton University
Campuzano, Juan Carlos	University of Illinois-Chicago
Car, Roberto	Princeton University
Cava, Robert	Princeton University
Chaikin, Paul	NYU
Chandra, Premi	Rutgers University
Chen, Mohan	Princeton University
Coleman, Piers	Rutgers University
Cooper, Lance	University of Illinois, Urbana-Champaign
Coppersmith, Susan	University of Wisconsin-Madison
Cross, Michael	Caltech
Deng, Xiaoyu	Rutgers University
Deshmukh, Amol	CUNY Graduate Center
Devarakonda, Aravind	Rutgers University

Dietrich, Scott	CUNY Graduate Center
Ding, Hong	Institute of Physics, Chinese Academy of Sciences
Dou, Wenjie	University of Pennsylvania
Doucot, Benoit	CNRS; University Pierre and Marie Curie
Drew, H. Dennis	University of Maryland
Drozdov, llya	Princeton University
Dziashko, Yury	UC Berkeley
Fang, Chen	Princeton University
Feldman, Ben	Princeton University
Fisch, Ronald	—
Fisher, Daniel	Stanford University
Fu, Ming-Xuan	McMaster University
Fu, Yaotian	Washington University
Genack, Azriel	Queens College of CUNY
Gershenson, Michael	Rutgers University
Goetschy, Arthur	Yale University
Graham, Robert	Universität Duisburg/Essen
Gubser, Steven	Princeton University
Gyenis, Andras	Princeton University
Haldane, F. Duncan M.	Princeton University
Halperin, Bertrand	Harvard University
Hamann, Donald	Rutgers University
Han, Qiang	Rutgers University
Han, Tian-Heng	University of Chicago and Argonne National Laboratory
Hasan, M. Zahid	Princeton University
Hohenberg, Pierre	New York University
Hopfield, John	Princeton University
Hristopulos, Dionissios	Technical University of Crete
Hsu, Ted	Member of Canadian Parliament
Huang, Biao	Ohio State University
Huse, David	Princeton University
Jeon, Sangjun	Princeton University
Johnson, Peter	Brookhaven National Laboratory
Kallus, Yoav	Princeton University
Kanter, Ido	Bar-Ilan University
Kharitonov, Maxim	Rutgers University
Klebanov, Igor	Princeton University
Kohn, Walter	UC Santa Barbara
Koirala, Nikesh	Rutgers University
Kosterlitz, J. Michael	Brown University

Kotliar, Gabriel	Rutgers University
Kulkarni, Manas	Princeton University
Kumar, Akshay	Princeton University
Langer, James	UC Santa Barbara
Lee, Patrick	MIT
Leggett, Anthony	University of Illinois
Leibler, Stanislas	Rockefeller University, Institute for Advanced Study
Levin, Kathryn	University of Chicago
Li, Lu	University of Michigan
Li, Yi	Princeton University
Libchaber, Albert	Rockefeller University
Lin, Chaney	Princeton University
Liu, Chang	Princeton University
Liu, Minhao	Princeton University
Luo Kai	CUNY Graduate Center and Hunter College
Maciejko, Joseph	Princeton University
Maguire, Camm	—
McDonald, Kirk	Princeton University
Metzner, Walter	Max Planck Institute for Solid State Research
Morel, Pierre	retired
Murray, James	National High Magnetic Field Lab
Nadj-Perge, Stevan	Princeton University
Nandkishore, Rahul	Princeton University
Nappi, Chiara	Princeton University
Neupert, Titus	Princeton University
Ogata, Masao	University of Tokyo
Ong, N. Phuan	Princeton University
Osheroff, Douglas	Stanford University
Osheroff, Phyllis	Stanford University
Pal, Arijeet	Harvard University
Papaioannou, Antonios	City University of New York
Petta, Jason	Princeton University
Poppe, George	CUNY Graduate Center
Ramakrishnan, Tiruppattur	Banaras Hindu University
Ramires, Aline	Rutgers University
Randeria, Malika	Princeton University
Randeria, Mohit	Ohio State University
Raval, Amita	Springer
Redi, Martha	Princeton Plasma Physics Laboratory
Ren, Yong	Saba Capital Management
Reyes Lillo, Sebastian	Rutgers University

Rice, Maurice	ETH Zurich
Ruderman, Malvin	Columbia University
Salehi, Maryam	Rutgers University
Sarachik, Myriam	City College of New York, CUNY
Sauls, James	Northwestern University
Schiro, Marco	Columbia University
Shastry, Sriram	University of California Santa Cruz
Shulman, Robert	Yale University
Stein, Daniel	New York University
Steinhardt, Paul	Princeton University
Talman, Jim	Western University
Tao, Rongjia	Temple University
Tchernyshyov, Oleg	Johns Hopkins University
Thouless, David	University of Washington
Tosatti, Erio	SISSA and ICTP, Trieste
Trivedi, Nandini	Ohio State University
Tsui, Daniel	Princeton University
Tsvelik, Alexei	Brookhaven National Laboratory
Tureci, Hakan	Princeton University
Vafek, Oskar	Florida State University
Vayl, Steven	City University of New York
Verlinde, Herman	Princeton University
Vollhardt, Dieter	University of Augsburg
Weeks, John	University of Maryland
Wheatley, Joseph	Biospherica Risk
White, Steven	UC Irvine
Wilczek, Frank	MIT
Witten, Edward	IAS, Princeton
Xiong, Jun	Princeton University
Yazdani, Ali	Princeton University
Ye, Meng	Rutgers University
Yin, Zhiping	Rutgers University
Youmans, Cody	CUNY
Yu, Clare	UC Irvine
Zandonella, Catherine	Princeton University, Office of Communication
Zhou, Brian	Princeton University
Zou, Zhou	ZZVENTURE

www.ingramcontent.com/pod-product-compliance
Lightning Source LLC
Chambersburg PA
CBHW061809210326
41599CB00034B/6941